MARINE CORPS WOMEN'S RESERVE IN WORLD WAR II

By

Lieutenant Colonel Pat Meid, USMCR

Printed 1964
Revised 1968

Historical Branch, G-3 Division
Headquarters, U. S. Marine Corps

1968

DEPARTMENT OF THE NAVY
HEADQUARTERS UNITED STATES MARINE CORPS
WASHINGTON, D. C. 20380

PREFACE

This brief history of the Marine Corps Women's Reserve in World War II is derived from official records and appropriate published and manuscript sources.

It is reprinted for the information of those interested in the wartime organization, training, activities, and record of service of women who served as Marines from 1943 - 1946.

R. G. OWENS, JR.
Brigadier General, U. S. Marine Corps
Assistant Chief of Staff, G-3

Reviewed and approved: 14 June 1968

TABLE OF CONTENTS

Foreword

Preface

Introduction. 1

I. Formation and Early History of the Women's Reserve. . 2

Preliminary Planning--Early Strength Estimates and Quotas--Selection of the Director

II. Public Announcement and Early Recruiting. 7

Public Announcement--Name vs. Nickname--The Enlistment Process

III. Training of the Women Reservists. 13

Officer Training--Recruit Training--Clothing Instructions--Transfer to New River--Troop Trains--"Hometown" Platoons--Training at Camp Lejeune--Specialist Training--Promotion from the Ranks--Reserve Officer Class

IV. The Uniform. 23

Official Issue--Those Dress Whites!--Special Uniform Class and Uniform Distribution--Uniform Board and Regulations

V. Jobs and Job Assignments. 28

Job Classification--Promotion--The buildup--Jobs in Aviation--"Appropriateness" of Jobs--The Philosophy of Hard Work

VI. Administration and Policies. 36

Cooperation with the Women's Services--Policy about Assignment and Housing Assistants for the Women's Reserve--Policy about Women's Authority--Changing Policy on Marriage--Discipline and Morale

VII. People in the Program. 42

The Women's Reserve Band--Quantico's Drill Team--Personalities in the WRs--Other WR Interests and Activities--Decorations Awarded Women Marines

VIII. Hawaii Duty. 48

Selection of Women for Overseas Duty--
Advance Party--Staging Area and Arrival

IX. Demobilization. 53

Computation of Credits--Strength at End of
the War--Monthly Quota for Demobilization--
Separation Centers--Last Days of the Wartime
Reserve

X. Overview. 58

Civilian Background--Reaction to Military
Life--Personal Benefit--Educational Back-
ground--Regional Pattern of Enlistments--
Composition of Reserve by Age and Test Scores--
Recruiting Results and Media Used--Overall
Distribution by Rank--"VIP" Statements about
Wartime Reserve

Notes. 69

Appendix A. Jobs in which Women Marines Were
Assigned During World War II. 88

Appendix B. Composition of the Women's Reserve:
By Education. 91

Appendix C. Composition of the Women's Reserve:
By State of Residence. 92

Appendix D. Composition of Women's Reserve:
By Age. 93

Appendix E. Composition of Women's Reserve:
By General Classification Test Scores. 94

Appendix F. Key Dates in the History of Women
Marines. 95

Appendix G. Biographies of Wartime Directors, Marine
Corps Women's Reserve. 100

MARINE CORPS WOMEN'S RESERVE IN WORLD WAR II

By

LtCol Pat Meid, USMCR

INTRODUCTION

"What! Women Marines! Quit your kidding."

That was the first reaction of a group of Marines newly-freed from a prison camp in the Philippines in February 1945. Eagerly they sought news from the combat correspondents about what had been going on in the Marine Corps since their capture in the early days of the war. The released men could hardly beleive it. Women in the Marine Corps? What did they do? How did they dress? What were they like? Were they pretty?(a)

Women in military uniform were a novelty to much of the rest of the world in the beginning of World War II, not only in this country, but in Canada and England as well. In the United States, more than 265,000 women served in all branches of the Army, Navy, Coast Guard, and Marines.(b) The Marine Corps Women's Reserve (MCWR) was established by law as a part of the Marine Corps Reserve by the amendment of 30 July 1942 to the U.S. Naval Reserve Act of 1938. The mission of the MCWR was to provide women trained and qualified for duty in the shore establishments of the Marine Corps, thereby releasing additional male Marines for combat duty.

In February 1943, the month that the Women's Reserve was formed, American forces wiped out the final enemy opposition on Guadalcanal. The bitter fighting there made it readily apparent that far more Marines would be needed in the combat zones for the grinding battles that would only slowly clear the way to victory.(c)

If the women caused innovations and creation of new traditions in the Corps, the effect of the Corps on them was no less profound. It was found that there are no differences between men and women in respect to their fierce pride in the Marine Corps and that special "Once a Marine, always a Marine" brand of loyalty. Years after the war, the story is told of a motorist on his way into the Marine Base at Quantico, who stopped at the Iwo Jima statue and picked up a little boy about nine years old. He glanced at the youngster sideways as he drove along, noticing that the boy wore a Marine Corps emblem on his cap.

"Is your father a Marine, son?" he asked conversationally.
"No sir," replied the boy. Then he added proudly, "But my mother was."(d)

This is the story of those World War II Women Marines--why they were there, the varied jobs they did, and their contribution to the war effort.

I. Formation and Early History of the Women's Reserve

On 7 November 1942, just three days before the 167th birthday of the Marine Corps, the Commandant signed a document that would bring about a great change in the life of the Corps during the years ahead. He gave his official approval to the formation of the Marine Corps Women's Reserve, a movement that resulted in more than 20,000 women serving in this ruggedly male outfit during the next three years and releasing urgently-needed male Marines for combat duty. Marine Corps Headquarters quickly and quietly went about setting up the policies and procedures needed to effect the innovation. Official announcement, however, was not made to the American public until three months later--on 13 February 1943. The Marine Corps had sought to avoid premature announcement before plans were completely developed, as it was felt this would not be in the best interests of the new Women's Reserve.

Actually, the Marine Corps was the last of the four services to organize a women's reserve in World War II.(1) This was no happenstance. It was generally well-known that the Commandant of the Marine Corps, Lieutenant General Thomas Holcomb, had been against the formation of a Women's Reserve in the Marine Corps at the time the Women's Army Auxiliary Corps (WAAC) and the Navy's Women Accepted for Volunteer Emergency Service (WAVES) were first organized. As he later commented, less than a year after the Women Marines had come aboard: "Like most other Marines, when the matter first came up I didn't believe women could serve any useful purpose in the Marine Corps....Since then I've changed my mind."(2)

Strangely enough, there had been a precedent for women in the Marine Corps. In World War I, a group of 305 intrepid young women had worn the forest green uniform with its famous globe and anchor insignia and had held private, corporal, and sergeant ratings. They had performed their clerical jobs, mostly at Headquarters, with dispatch and loyalty and had served under the traditional no-nonsense Marine Corps discipline which decreed that any infractions on their part would result in their being "summarily disenrolled."(3) They had even been instructed in the "simpler drill movements...before nine o'clock on the ellipse in Potomac Park," under the watchful eye of a Marine NCO, and had also participated in victory parades.(4)

Rightly, however, when the matter of a Women's Reserve was discussed as a serious possibility in the Second World War, General Holcomb knew that admission of Women Marines this time would be on a scale and magnitude previously undreamed of. It would create hundreds of new questions and problems, all of which had to be answered.

Preliminary Planning

The decision to admit women to the Marine Corps was made as a result of studies prepared by the M-1 section of the Division of Plans and Policies at Marine Corps Headquarters. Originally, the matter had been discussed months earlier but, because of the Commandant's feeling, had been dropped. When it became apparent that such a move would release large numbers of urgently-needed combat personnel, the question was reopened and restudied.

On 5 October 1942, Plans and Policies recommended establishment of a Women's Reserve and suggested to the Commandant that it be set up as a separate section within the Division of Reserve.(5)

The Commandant concurred with the recommendation, and on 12 October wrote the Secretary of the Navy that "in furtherance of the war effort, it was believed that as many women as possible should be used in noncombatant billets, thus releasing a greater number of the limited manpower available for essential combat duty."(6) He cited Public Law 689, 77th Congress, approved 30 July 1942, which amended the Naval Reserve Act of 1938 by adding a section titled "Women's Reserve" and provided that it should be part of the Naval Reserve.

First endorsement of the Commandant's letter, on 26 October by the Judge Advocate General's office, approved this legal authority and read: "The creation of a Women's Reserve which shall be a branch of the United States Marine Corps Reserve appears to be fully authorized by the law. The specific proposals of the Commandant of the Marine Corps as contained in the basic communication have been examined in this office and it is considered that they legally be approved."(7)

The second endorsement on 30 October from the Commander in Chief, United States Fleet and Chief of Naval Operations was similarly favorable. "Forwarded, recommending approval," it read in time-honored official naval language.(8)

Final authority for creation of the Women's Reserve was received from Secretary of the Navy Frank Knox on 31 October and President Franklin D. Roosevelt on 7 November. This authorized an initial strength of 500 officers and 6,000 enlisted by 30 June 1943, with total strength by 30 June 1944 of 1,000 officers and 18,000 enlisted.(9)

Distribution of rank and grade was the same as that authorized for the men of the Marine Corps. Based on the number of women to be enlisted, the distribution in rank of officers was specified as: 1 major, 35 captains, 35 percent of the total number of commissioned officers to be in the grade of first lieutenant, and the balance, to be second lieutenants.(10) The determination of the highest rank to be held by a member of the Marine Corps Women's Reserve was based on the language of Public Law 689, which provided for one officer with the grade of lieutenant commander for the Women's Reserve of the United States Naval Reserve, whose counterpart in the Marine Corps would hold the rank of major.(11) Later amendments to the law advanced the rank of the senior woman in each naval service to captain (Navy and Coast Guard) - colonel (Marine Corps).

Early Strength Estimates and Quotas

The first definite step toward physically establishing the Women's Reserve was taken 5 November, when the Commandant wrote the commanding officers of all Marine posts and procurement districts. He announced that the Marine Corps was "initiating steps to organize a Women's Reserve" and directed all officers to survey the activities under their jurisdiction and report the number of Women Reservists (WRs) who could be used to replace officers and men in such categories as clerical, communications, transportation, mess and commissary, mechanical, and so forth. He explained that "within the next year the manpower shortage will be such that it will be incumbent on all concerned with the national welfare to replace men by women in all possible positions."(12)

From the figures submitted by posts and stations, projections were made of the number of women who would require special training in such fields as Paymaster, Quartermaster, and Communications, as well as estimates of those WRs who could effectively use the skills they brought to the Marine Corps from civilian life. Quotas were also established for recruiting of enlisted women and officers and tentative dates selected for the beginning training classes.(13)

Although the Marine Corps was authorized a strength of only 6,500 women by 30 June 1943, this preliminary survey indicated that more than 4,000 were needed <u>at once</u>. The number of WRs originally requested at Marine Corps stations were:

Quantico, Virginia	692
Cherry Point, North Carolina	688
Camp Lejeune, New River, North Carolina	726
San Diego, California	650
Camp Elliott, California	981
Camp Pendleton, California	416
Parris Island, South Carolina	278
	4,431 (14)

Selection of the Director

Considerable preliminary planning had to be done to facilitate successful recruiting, training, administration, and uniforming of the new Women's Reserve. But probably the most important task confronting the Division of Reserve was selection of a director for the new reserve component. The Commandant made no bones about the fact that the success of the new Women's Reserve would depend largely on the caliber and capabilities of the woman chosen for the post of Director, Marine Corps Women's Reserve. Accordingly, in November he wrote Dean Virginia C. Gildersleeve, of Barnard College, Columbia University, to enlist her help.

"It is my understanding that in the selection of the woman to head up the WAVES, the Navy availed itself of the advice of the Advisory Educational Council, of which you are chairman," he wrote. "If it is not too much of an imposition, the Marine Corps would be glad if your council could undertake a similar service for it." General Holcomb stressed the point that it was not the intention of the Marine Corps to dictate any method in the choice of candidates--whether decided upon by subcommittee or other means. "We are only interested in procuring the services of some woman who is qualified for a commission as a Major in the Marine Corps and to assume the parallel position to Miss McAfee /Lieutenant Commander Mildred McAfee, Director of the WAVES/," he explained. (15)

Shortly thereafter, Dean Gildersleeve and her committee presented a recommendation of 12 outstanding women, and the Marine Corps began making discreet inquiries as to their capabilities. Personal interviews of the various candidates were conducted by Colonel Littleton W. T. Waller, Jr., Director of Reserve, under whose office the new Women's Reserve was to be placed for administrative purposes. Colonel Waller and his right-hand man, Major C. Brewster Rhoads, toured the country to interview prospective candidates personally. Their recommendations ultimately led to selection of Mrs. Ruth Cheney Streeter, 47, of Morristown, New Jersey.

President of her class at Bryn Mawr College; the mother of four grown children, including three service sons (two in the Navy, one in the Army); for more than 20 years active in New Jersey health and welfare work; and a spirited woman who only a year or so earlier had taken out both her private and commercial pilot's licenses; Mrs. Streeter seemed to have the right combination of personal characteristics and organizational abilities that would be required of a Woman Marine Director.(16)

Prior to public announcement of the new Women's Reserve, Mrs. Streeter was quietly commissioned a major, USMCWR, on 29 January 1943 and sworn in by Secretary of the Navy Knox. She was not, however, the first woman to go on active duty in the Women's Reserve in World War II. Earlier that month, Mrs. Anne A. Lentz was sworn in as the first commissioned officer, with the rank of captain. (A civilian clothing expert who had helped outfit the WAACs, she had originally come to the Marine Corps Headquarters in December on a 30-day assignment to design the uniform for the Marines and wound up by wearing one herself.)(17)

In respect to procurement and training, existing facilities of the WAVES were to be used as much as possible. This was spelled out in a joint letter from the Chief of Naval Personnel and the Commandants of the Marine Corps and the Coast Guard to the Secretary of the Navy. Since the Women's Reserve of both the Marine Corps and Coast Guard were also part of the Naval Establishment, it was officially recommended that their members be procured through the Office of Naval Officer Procurement and be trained "insofar as it is practical" in schools already established for members of the WAVES.(18)

In February 1943 there were six other women who, like Major Streeter and Captain Lentz, had been directly commissioned from civilian life before actual public announcement of the Women's Reserve. All were selected because their abilities and experience fitted them for key Marine Corps billets which had to be filled at once--such as recruiting and training. Commissioned without any formal indoctrination, they went on duty immediately at Marine Corps Headquarters with their new rank and in civilian clothes.

These early Women Marines were:

Women's Reserve representative for Public Relations--First Lieutenant E. Louise Stewart;

Women's Reserve representative for Training--Captain Charlotte D. Gower;

Women's Reserve representative for Classification and Detail--Captain Cornelia D. T. Williams;

Women's Reserve representative for West Coast Activities--Captain Lillian O'Malley Daly;

Women's Reserve representative for Recruit Depot--Captain Katherine A. Towle; and

Women's Reserve Assistant to the Director--Captain Helen C. O'Neill.(19)

II. Public Announcement and Early Recruiting

Although the last-organized of the four women's wartime services, three important factors were in the Women Marines' favor from the start.

First: That the Marines freely shared their own name, a proud name that had witnessed 168 years of tradition and esprit. Thus, they became the only women's service which didn't have an alphabetical designation or semi-official nickname.

Second: That the Women's Reserve was accepted as a full-fleged part of the Marine Corps and was not an "auxiliary" service.

Third: That the men's distinctive forest green uniform was followed closely, with requisite feminizing modifications for the Women Marines. This, like the name, made the women feel they were being accepted on an equal basis in the Corps, rather than as an auxiliary, and they worked twice as hard to make sure they rated being called "good Marines."

Colonel Waller, Director of Reserve, in a recommendation to the Commandant a month before announcement of the Women's Reserve was made to the public, declared:

"Women Reservists of the U. S. Marine Corps will not be especially designated as in the case of "WAVES" or "SPARS," but will be called Marines. It is proposed that they will be uniformed in the forest green of the Marine Corps with suitable differences being made in the material and in the cut of the uniform to conform to the convenience and smart appearance of women, but sufficiently like the Marine Corps uniform to permit no possibility of doubt as to the branch of service to which the Women Reservists are attached."(20)

Public Announcement

The first official announcement of the Women's Reserve was made on 13 February 1943, and the Navy procurement offices throughout the country which were charged with the duty of enlistment suddenly found themselves swamped with women who wanted to be Marines. In the nation's capital, more than 100 filed applications the first two days after enlistments opened and caused one recruiting officer to complain that the overload of applicants was causing his office staff to "get behind in their work."(21) Applicants ranged all the way from Mrs. Otho L. Rogers, of Washington, D. C., and Mrs. Henry T. Elrod, of Coronado, California, both widows of Marine majors recently killed in combat, to schoolgirls, office workers, grandmothers, and college students.(22)

Enthusiasm ran so high that a number of women even tried to enlist on the Saturday the initial announcement was made, even though enlistments were not supposed to be officially open until the following Monday. The records shows that some of them succeeded. The distinction of being the first World War II Woman Marine (other than the handful of officers direct-commissioned before public announcement) went to Lucille E. McClarren, of Nemahcolin, Pennsylvania, who enlisted in Washington, D. C. on 13 February.(23)

Eligibility requirements for both enlisted and officers were:

United States citizenship; not married to a Marine, either single or married but with no children under 18; height - not less than 60 inches; weight - not less than 95 pounds; good vision and teeth.(24)

For enlisted or "general service," as it was called, the age requirement was from 20 to 35 inclusive, and a candidate was required to have at least two years of high school.(25)

For officer candidates, requirements were originally the same as for WAVES and SPARS: age from 20 to 49 inclusive; either a college graduate, or with a combination of two years of college and two years of work experience.(26)

From the very beginning, it was a problem for the Marine Corps to cope adequately with the stream of volunteers. Through courtesy of the WAVES, the Navy Department made a unique and generous offer: some of its own officers, currently undergoing training, would transfer to the Marine Corps to help with recruiting, if the Marine Corps so desired. The Marine Corps, sorely pressed for personnel, was happy to reply "Yes!" A number of WAVES volunteered, and a group of 19 was selected, since the Marine Corps had 19 procurement offices throughout the country. These 19 ex-WAVES were sworn in as new Marines and went on the job immediately to recruit Women Marines. Ironically, they still wore their WAVE uniforms, as the first Marine Corps uniforms were not yet available.(27)

Throughout those early hectic months there was, inevitable, much "trial by error" and the type of resourceful improvisation that has always been the hallmark of the Marine Corps. Some Marine officers, for example, in the large cities, who were severely pressed for additional help in the mountain of paperwork inherent in enlisting large numbers of women, wrote to Headquarters requesting authority to enlist capable civilian women as Marines to work in their own offices. The women would go on duty immediately to help out in the critical overload of work, and would receive their actual indoctrination as Women Marines later. These requests were granted in many cases.(28)

Minimum age for a prospective member in the Women's Reserve had been set by Congress in its 1942 amendment to the Naval Reserve Law and remained unchanged throughout the war. It was, nevertheless, a matter which aroused considerable agitation on the part of younger women, and both the White House and the Director received numerous letters on the subject. In one instance, an articulate young lady, representing a group of nearly twenty 18- and 19-year-old girls from Springfield, Ohio, wrote a highly-persuasive letter to Major Streeter, asking the logic in a ruling which permitted 18-year-old boys to defend their country, even at the supreme sacrifice, when the girls could not. Another enterprising young woman from Avonmore, Pennsylvania, who obviously was well-versed in her facts, pointed out that in the last war girls 18- and 19-years old were allowed to enlist in the Marines and girls not quite 18 could join with their parents' consent. "If girls 17 were allowed to enroll in the last war," she asked Mrs. Roosevelt, "could it be possible for a girl 19 to enroll in the Marines today?"(29)

A number of parents also wrote to ask if their daughters could enlist, even though not yet the required age of 20. One such request came from a World War I holder of the Distinguished Service Cross. In January 1943, prior to actual formation of the Women's Reserve, he wrote the Commandant:

"I know this is no time to reminisce, but I do want to bring this to your attention. I am the Marine from 96th Company, Sixth Regiment, who was with Lieutenant /Clifton B./ Cates and a few other Marines that captured Bouresches, France, and I turned over the first German prisoner and machine gun to you that our battalion captured on the night of 6 June 1918.

"I have a big request to ask....As I have no sons to give to the Marines, I would be more than happy if you....would recommend my daughter to the newly-formed Marines Women Reserve Corps. While I appreciate that her age may be a little young, she will be 18 this June....I feel sure she could fit into your program....surely this is not too much for a D. S. C. ex-Marine to ask of you...."(30)

The minimum age limit of 20 years for women members of the Naval Reserve had been established by law, however, and so it remained. Persons writing such letters were thanked for their interest in the Marine Corps, with the suggestion being made that the girl reapply later when she became the proper age.

Name vs. Nickname

The public took great interest in every detail about the new organization and freely submitted trick names for the new Women's Reserve even before the time it was officially formed. Unsolicited suggestions came from Congressmen and private citizen alike. Typical "names" included MARS, Femarines, even the unwieldy Women's Leather-neck Aides.(31)

However, the firm decision had been made that the Women Reservists would be called simply "Marines." As the Director of Reserve, Colonel Waller, wrote to Representative Louis Ludlow of Indiana: "...these women will not be auxiliary but members of the Marine Corps Reserve which is an integral part of the Corps and as....they will be performing many duties of Marines it was felt they should be so known."(32)

The Enlistment Process

Candidates both for officer and recruit (boot) training were enlisted by naval procurement offices in each of the four major procurement districts into which the country was divided. New enlistees were placed on "inactive duty" while their applications and supporting papers were processed by Marine Corps Headquarters and then notified of the training class, whether boot or officer, to which they were assigned. The women were designated as either Class VI (a) officer or Class VI (b) enlisted. (Previously established male categories of the Reserve were: Class I - Fleet Marine Corps Reserve; Class II - Organized Marine Corps Reserve; Class III - Volunteer Marine Corps Reserve; Class IV - Limited Service Marine Corps Reserve; and Class V - Specialist Volunteer Marine Corps Reserve.)(33)

A prospective applicant, regardless of classification, had to submit a physical statement by her own physician; fill out and return the application to the recruiting station; take an aptitude test and complete physical; and have a personal interview with the Officer-in-Charge. He in turn submitted a statement of opinion as to the woman's capabilities and value to the service. Final decision on whether or not to accept a particular candidate was made by Headquarters Marine Corps. In some cases, an otherwise well-qualified candidate was permitted a waiver for a physical or educational requirement she lacked.

The traditional selectivity of the Marine Corps was in evidence at once. Indication of its high standards is seen in the fact that the Marine Corps Reserve Reviewing Board at Headquarters rejected approximately 25 percent of the applicants for officers candidates' class whose applications were forwarded from the procurement offices.(34) Public response to recruiting was eminently satisfactory. Summarizing the first month's progress of the new women's service, Colonel Waller commented:

"The women of the country have responded in just the manner we expected....Thousands of women have volunteered to serve in the Women's Reserve and from them we have already selected more than 1,000 for the enlisted ranks and over 100 as officers."(35)

This was a good record when it is remembered that the WR goal was only 6,000 enlisted and 500 officers by 1 July, the maximum number that could be enrolled up to that time.

Some of the interest of the nation's women was undoubtedly stirred by the nationwide trip which Major Streeter made during the first month of the Women's Reserve. She visited 16 major cities from coast to coast, as well as the Marine Corps posts where her women would shortly be serving. She spoke before a number of large public gatherings, including women's clubs and college assemblies, and everywhere found a "spontaneous enthusiasm among women for the new women's service organization."(36) Indeed, so ambitious was her schedule of speaking engagements that at one point her voice gave out!(37) Returning to Washington on 26 March from the first of what would be many trips to the field, she commented: "The privilege of swearing in many enlistees gave me an opportunity to observe the young women joining our ranks. I found them to be most sincere, intelligent, and attractive representatives of American young womanhood."(38)

Enlistments in the Women's Reserve during its first eight weeks totaled 2,495. Of that number, 28 were on active duty; 211 were enrolled in officer candidates' class; and the rest were either in recruit training or awaiting orders to active duty. During this first two months of its existence, nearly one-seventh of the total enlisted strength and more than one-quarter of the future officers were enrolled.(39)

In at least one place the recruiting was going so well that it occasioned this cautioning remark from the procurement officer-in-charge: "While it is not properly a concern on this office, it is felt that the Division of Reserve might well consider decreasing the overall publicity given to the women's program since the number of applications far exceeds the authorized quotas. It is considered to be bad public relations by this office when unavoidable circumstances necessitate turning away many desirable and well-qualified applicants who have been encouraged to believe that their enlistment could be effected."(40)

Despite the problems inherent in such an undertaking, by July 1943 sufficient female personnel had been trained so that it was possible to transfer all enlistment procedures from the naval procurement offices to Marine recruiters, who from then on handled the enrollment of the women as they had been doing right along in the case of the men. As always, physical examinations were the responsibility of the Navy.(41)

By 1 November 1943, the number of officers and enlisted personnel sworn into the service totaled more than 11,000--less than 1,000 short of the 12,000 member quota set for 1 January 1944. Of this number, approximately 8,500 had been classified and were on duty. In February 1944, one short year after its formation, the strength of the Women's Reserve totaled nearly 15,000. A year earlier, the organization had consisted of four officers; now, it numbered approximately 800 officers and 14,000 enlisted women, and was well within sight of its final recruiting goal.(42)

Observed the Director in May 1944: "...it is anticipated that by 31 May the Marine Corps Women's Reserve will have reached its total authorized strength. Although our quota is much smaller than those of the WACs (Women's Army Corps) or WAVES, the fact still remains that though we were the last of the women's military services to be organized, we are the first to succeed in enlisting all the women we can presently use. Furthermore, we reached our goal in two and a half months less time than we expected. The original plan called for recruiting to start on 1 January 1943 and proceed at the rate of 1,000 per month until 30 June 1944. Actually, we did not start until 15 February 1943 and reached our quota on 31 May 1944, thus accomplishing our mission in 15½ months instead of 18."(43)

III. Training of the Women Reservists

In addition to its assistance with recruiting, the Navy also offered the use of its training facilities for both officers and enlisted personnel in the early organizational stages of the Women's Reserve. This helped the Women Marines get off to a good start. Had it been necessary for the Marine Corps to train its own staff for its women's schools before they could begin to operate, there would have been a delay of several months in putting the women to work for the Marine Corps and releasing men for combat duty. Classes for officers and enlisted women both began in March 1943, at Mount Holyoke College and Hunter College, respectively.

Officer Training

On 13 March 1943, exactly a month after initial public announcement of the Woman's Reserve, the first class of 71 officer candidates entered the U. S. Naval Midshipmen's School (WR), Northampton, Massachusetts, to begin its training with the WAVES. The U. S. Naval Midshipmen's School comprised the facilities of both Smith College, at Northampton, and Mount Holyoke, in nearby South Hadley. Marine candidates received their training primarily at the latter.(44) Included in the class were representatives of a variety of fields of civilian life--including educators, scientists, secretaries, and women from other businesses and professions.(45)

Marine officer candidates followed the same course of instruction as the WAVES for the first half of their training, approximately four weeks. This included Naval Organization and Administration, Naval Personnel, Naval History and Strategy, Naval Law and Justice, Ships and Aircraft. Instruction in the second part, or advanced indoctrination, was separate from the WAVES. This consisted of Marine Corps subjects given by Marine Corps instuctors. The curriculum included Marine Corps Administration and Courtesies; Map Reading; Interior Guard; Safeguarding Military Information; and Physical Training. As with all Marine recruits, throughout both phases of the entire course, the women were schooled rigorously by male Marine drill instructors, who had been transferred from the Parris Island Recruit Depot to Mount Holyoke for this purpose.(46)

On 6 April, members of the first class received their silver OC pins, which marked their promotion to officer cadet status. This was a specially-created category, authorized by the Secretary of the Navy, to correspond to the status and pay rating of their contemporaries in the WAVES. Since Navy candidates went through their training as midshipmen, it was felt desirable for members of the two groups to have equal standing.(47) Successful candidates received their commissions on 4 May, a little over seven weeks after they had entered. The second candidates' class began its training on 10 April and the third class, in early May. Altogether a total of 214 Women Marines completed officer training at Mount Holyoke, with a new class entering each month.(48)

Administratively, the Marine training unit had the comparable status of a Marine detachment aboard ship. The women had their own commanding officer who was responsible for discipline, as well as for coordinating instruction in drill and academic subjects with that of the school. The Marine candidates were organized into separate companies and were under the immediate command of an officer of the regular Marine Corps, Major E. Hunter Hurst. But the WR detachment itself was part of the WAVES school complement, under final authority of the commanding officer of the Midshipmen's School.(49)

All officer candidates enlisted as privates. At the end of their preliminary four-week training period, women considered not qualified for appointment as cadets had the option of either being transferred to Hunter College for completion of basic training, or of being ordered to their homes and inactive status in the reserve district to wait ultimate discharge from the Marine Corps. Cadets who, upon completion of their training, were not recommended for commissioned rank, submitted their resignations to the Commandant via official channels and were subsequently discharged. If they wanted to reenlist as a private, they could do so, provided they were not over age for enlistment.(50)

Recruit Training

Two weeks after the first officer class began its training, the first class of 722 enlisted women entered Hunter College, The Bronx, New York or, as it was officially known, the U. S. Naval Training School (WR). Due to the size of the group, its members were ordered to arrive over a three-day period, 24-26 March, in three equal daily contingents.(51) The "boots" were billeted in nearby apartment houses and began their instruction with the WAVES on 26 March.

The administrative set-up was similar to Mount Holyoke but vastly larger. Although part of the host Navy organization, the Marine recruits were organized into separate companies, each headed by a male Marine officer and combined into a battalion, under command of an officer of the regular Marine Corps, Major William W. Buchanan. The first class was divided into 21 platoons of approximately 35 women each.(52)

A senior woman officer, Captain Katharine A. Towle, was a member of Major Buchanan's staff from the beginning. Other women officers were added to it after the first officer candidates' class was commissioned. A group of 33 instructors, including 10 officers and 23 enlisted, comprised the major's staff. They instructed the women in both Marine Corps and general subjects, the curriculum being similar to that of Mount Holyoke. In addition, there were 15 to 20 sharp-eyed drill instructors to supervise the close-order drill of all the women in the training school, both WAVES and Marines.(53)

Included in the first group of enlisted women were many stenographers and secretaries, telephone operators, two motor mechanics, laboratory technicians, an acetylene welder, a commercial artist, a parachute maker, woodcraft workers, and others representing a wide variety of occupations and civilian skills.(54)

The first class was graduated 25 April 1943, in a little over four weeks' time. Subsequent classes entered every two weeks, and numbered some 525 recruits each. Indoctrination lasted approximately four weeks, but individual classes varied from three-and-a-half to five weeks because of the need to coordinate schedules with the WAVES. Between 26 March and 10 July, six recruit classes entered and a total of 3,280 women Marines were graduated. Despite the intensity and fast pace of the training, attrition--about two percent--was quite low.(55)

Clothing Instructions

Members of both the first candidates' class and recruit class went through half their training in civilian clothes. Uniforms were issued in the latter part of April, as soon as they became available in quantity. Detailed instructions issued to prospective WRs before they left home for training had spelled out clearly the clothes they should bring with them, encluding two pair of comfortable dark brown, laced oxfords. "Experience has proven that drilling tends to enlarge the feet," the mimeographed instructions stated matter-of-factly. In addition to the list of necessary clothing, all trainees were sharply warned not to leave home without orders; not to arrive before the exact time and date stamped on their official papers; and not to forget their ration cards.(56)

Transfer to New River

By the early summer of 1943, the Marine Corps had readied its own schools. Although it was originally under the orders to use existing facilities of the Navy insofar as was practical for procurement and training, the size of the classes both at Hunter and Mount Holyoke dictated the need for larger facilities.

In July 1943, the fifth month of the Women's Reserve, both the officer candidates' class and recruit depot were transferred to Camp Lejeune, New River, North Carolina. Together with the specialists' schools, which had been in operation at New River since May, they comprised the Marine Corps Women's Reserve Schools. Here, nearly 19,000 women took their training throughout the remainder of the war.

The third class of officer candidates was commissioned at Mount Holyoke on 29 June. The combined battalion of WAVES and Marines passed in review before Major Streeter, Lieutenant Commander McAfee, Director of the WAVES, and Brigadier General Keller E. Rockey, USMC, Director of the Division of Plans and Policies at Headquarters. Commented Major Streeter: "The candidates presented an excellent battalion review conducted entirely by themselves without any men officers on the field. They made a very good impression in all ways and left Mount Holyoke with good feeling between themselves and the Navy and the college."(57)

Possibly the satisfactory experience at Mount Holyoke was due partly to another factor. Its president was an ex-Marine! As the Marine Corps later wrote in a letter of appreciation to Dr. Roswell G. Hamm: "Your continual willingness to assist in the formation of policies and to contribute to the comfort of the Marine Corps personnel at Mount Holyoke were largely responsible for the high morale and fine <u>esprit de corps</u> of our officer candidates. Your experience as a former Marine made you keenly aware of the vital importance of the work to be done by the Women's Reserve."(58)

Tuesday, 29 June, was also the day that members of the fourth class were promoted to rank of cadet. On Thursday, approximately 70 members of the training class and the staff departed in a troop movement to Camp Lejeune, arriving two days later. Training was resumed on 5 July and the class graduated on 7 August. The fifth class reported directly to Camp Lejeune on 15 July as did all candidates' classes thereafter.(59) Meanwhile, at Hunter College, the current class of enlisted women completed its training in early July. The tenth class reported directly to Camp Lejeune on 12 July and graduated on 15 August. Thereafter, a class of approximately 550 women entered every two weeks and graduated about five and a half weeks later, in accordance with previously established schedules. Three classes were in training simultaneously.(60)

The new location and consolidation of training was welcomed by all, students and administrative staff alike. It enabled a far more thorough Marine Corps indoctrination than had been possible before and permitted later classes of enlisted women to receive detailed instructions in various administrative procedures needed on their day-to-day jobs.

A highlight of all Women Marines' training, initiated after the move to New River, were the field demonstrations in which the women witnessed actual use of mortars, bazookas, flamethrowers, amphibian tractors, landing craft, hand-to-hand combat, camouflage, even war dogs. Picked teams of male Marines presented these special demonstrations in half-day sessions. "By showing the women what the men faced whom they had released for combat, their pride in the Corps was increased and they saw clearly their own part in it," the Director of the Women's Reserve later observed. (62) Since no other women's military service had such real-life battle demonstrations, it was understandable that their members were somewhat envious of this aspect of the WR's training!(63)

Actually, the first tentative step toward what was later to become the field demonstrations had occurred informally less than a month after training began. A personal letter received by recently promoted Brigadier General Waller from Major Hurst, Commanding Officer of the Marine Training Detachment at the Midshipmen's School in South Hadley had stated, in part:

"In drawing these up /training schedules ordered by Marine Corps Headquarters/ I found myself wishing more and more that we could include some weapons instructions, at least pistol, for our women....I have found that the women come into the Marine Corps expecting to learn to shoot and I, of course, would like to see them become the first women's reserve in the country to take up the specialty of their men if Headquarters considers the idea at all feasible. I wouldn't have had the nerve to suggest it if Mrs. Franklin D. Roosevelt hadn't asked me on her visit last week how soon they were going to learn to shoot. She expressed surprise at learning that the women of the U. S. were not learning as much about weapons as the women of other countries...."(64)

Weapons demonstrations took another big step forward in a memorandum drafted 12 June 1943 by Major Streeter discussing the proposed curricula for the Marine Corps Women's Reserve Schools to open the following month in New River. She noted that the indoctrination of both training classes of women contained lectures on combat equipment, landing operations, tactics, parachute troops, and amphibian tractors. "If it is possible to arrange transportation and schedules that would not interrupt the training of the men in these lines of work, I believe it would be a definite inspiration to the Marine Corps Women's Reserve to see them actually in training," she wrote.(65) The Director's suggestion was approved, and the modification in the women's training was considered to be highly advantageous by all concerned.

Troop Trains

Lessons in Marine Corps style efficiency and order were learned even before new recruits arrived at Camp Lejeune. They were brought to New River on all-Marine trains---all-Women-Marine trains. Numbering approximately 500 girls, the mass troop movement was directed by a woman lieutenant, with two enlisted women as assistants. Commented one recruit: "We started right out learning military procedure and discipline at the railroad station. The WRs lined us up, bag and baggage, and marched us aboard the train."(66)

Once at Camp Lejeune, boots observed the strict rules governing male recruits at the Parris Island and San Diego boot camps. Every minute of the day was accounted for, and no liberty was granted during the six-week indoctrination.(67) Training got underway the minute the women arrived. Speedy assignment to billets in the neat red brick barracks in Area One, set aside for the exclusive use of the women's schools, was followed by orientation classes; issue of uniforms; close order drill, beginning the day after arrival; and classification tests and interviews to assess a woman's abilities, education, training, and business experience. Strict discipline and tight schedules worked their invariable magic. Before long it seemed a perfectly normal routine to get up at 0545, fall in formation at 0630, eat at 0645, attend classes from 0800 to 1130, march to lunch, and spend until 1600 daily in classes or drill.

Despite constant emphasis on discipline, proper military phraseology and customs, even the best-intentioned WR sometimes made mistakes, often ingenuous ones with a decidely feminine twist. There's the story of the Woman Marine who became flustered upon passing an officer on the street and got her instructions mixed. Instead of saluting and saying, "By your leave, sir," she saluted and said "Leave me by, sir."(68)

In another instance, a woman student platoon leader tried in vain to give her marching troops the order of execution on the correct foot, With her platoon marching along, she decided to compose herself for a minute to make double sure. Suddenly, dead ahead of the column, a tree loomed up. Her command rang out strong and clear: "Around the tree....MARCH!"(69)

"Hometown" Platoons

A month after the organization of the Women's Reserve, an officer in the Southern Recruiting District had queried Headquarters: "...we are making plans for the formation of a platoon of Women Marines to be sworn in jointly and sent to training as a group. This has been done successfully with male Marines in the past. If there is any objection to this, please wire immediately."(70)

Not only was there no objection from Headquarters, but the idea was picked up by other recruiting officials and cities. Although Atlanta appears to have been the city where the idea was originally conceived for the hometown platoons, it was the city of Philadelphia, birthplace of the Marine Corps back in the days of the Revolution, that produced the first WR platoon to be sent to camp as an entity. This occurred in early September 1943 and rated a telegram of congratulations and "Welcome Aboard" from Major Streeter.(71)

The 168th observance of the Marine Corps birthday on 10 November was the occasion for the swearing in en masse of both the first Pittsburgh platoon as well as the Potomac Platoon of Women Marines of Washington, D. C. The latter ceremony took place at the Library of Congress, prior to departure of the unit for boot training at Camp Lejeune. Much local enthusiasm was created, and, as a library official later wrote to a Marine officer: "In all my years of association with the Library of Congress I have never seen the steps of the main building put to more appropriate use than the swearing in of the First Potomac Platoon of Women Marines."(72)

The recruiting of other "all neighbor" women's platoons was scheduled which included: Albany, Buffalo (two), Northern New England, Pittsburgh (two), Miami, Alabama, Fayette County, Pa., Johnstown, Pa., St. Paul, Green Bay, Westmoreland County, Pa., Seattle, Houston, Southern New England, Central New York, and Dallas. Members of the platoons were ordered to duty at one time, went through their preliminary training as a unit, but upon completion of their training were assigned to duty individually.(73)

Training at Camp Lejeune

The training program of all Women Marines was drawn up with the prime objective of converting civilians into responsible military personnel in the shortest time possible. As with generations of male Marines before them, close-order drill proved to be the most effective single training factor. (74) Through these basic military movements, the Women Marines learned not only the value of teamwork, military precision and snap, instantaneous response to command, and discipline and order, but also pride in outfit, pride in self, and the intangibles of that traditional Marine esprit.

Upon completion of basic training, those women considered to have sufficient skills to be of immediate value to the Marine Corps received their orders and went on active duty at once. Other women, both officer and enlisted, were assigned to specialist schools and still others were trained on the job as apprentices. Depending on their civilian background and skill, some took over in supervisory positions. In any event, the aim was to get the women assigned to suitable duty as rapidly as possible in accordance with the current needs of the service.

Lieutenant Colonel Lucian C. Whitaker, USMCR, and, later Colonel John M. Arthur, USMC, were the commanding officers of the Camp Lejeune Women's Reserve Schools which included the Recruit Depot, Candidates' Class, and Specialist Schools Detachment. The platoons in training at New River averaged from 28 to 30 women, and a company, approximately 165.(75)

Of the 22,1999 women ordered to Recruit Depot (i.e., Hunter College and Camp Lejeune), only 602 failed to complete the course for physical reasons or inaptitude, an attrition rate of 2.7 percent. These individuals were discharged either on grounds of unsuitability or by medical survey.(76)

Specialist Training

From the very beginning, advanced training was available. More than 100 members of the first class graduated 25 April from Hunter attended Navy and Marine specialist schools. The early Navy courses were: Aviation Machinist Mate at the Naval Training School, Memphis, Tennessee; Link Training Instructor at the Naval Air Station, Atlanta, Georgia; and Aviation Storekeeper at Indiana University, Bloomington, Indiana. Early Marine specialty schools included cooks and bakers, motor transport, quartermaster, and non-commissioned officers. Members of the officer classes at Mount Holyoke were also selected for further training, including instruction at the Navy's communications school in South Hadley, Massachusetts.(77)

Before the war was over, some 30 specialist schools were open to the Women Marines in fields as diverse as mechanics and personnel administration. Nearly 9,000 women received such advanced training. The courses varied in length from 4 to 22 weeks and were open to women who had finished "boot training" and who sought and qualified for higher ratings in the specialized fields.(78) As women proved their versatility on the job, the original half-dozen or so specialist schools quickly expanded to capitalize on their abilities. At the University of Wisconsin, for instance, one WR studying radio communications actually picked up an SOS from a ship sinking somewhere at sea.(79)

In addition to the early schools, other Marine Corps and Navy training courses open to the women during the two and a half years of the war included: first sergeant, paymaster, signal, parachute rigger, aerographer, clerical, control tower operator, aerial gunnery instructor, celestial navigation, motion picture operators technician, aircraft instruments, radio operator, radio material, radio material teletypewriter, post exchange, uniform shop, aviation storekeeping, automotive mechanic, carburetor and ignition, aviation supply, and photography. Many of these classes, such as those at the First Sergeant's School in Philadelphia, contained old-time veteran Marines. Thus members of the new Women's Reserve benefited both by personal association with these highly competent (and sometimes highly critical) "Old Salts" and from classroom discussions of their job experiences. Top-ranking students were often awarded a higher rating than the majority of the class upon completion of specialty training.(80)

Promotion from the Ranks

The first seven officer candidates' classes were made up of women who enlisted in Class VI(a) directly from civilian life.(81) The applications of these women were forwarded to Headquarters from the procurement district where they had originally enlisted. In Washington, a four-member board reviewed all applications for officer training and selected the best qualified, who were subsequently ordered to duty.

Because there were many outstanding enlisted women who, officials believed, should also have the opportunity for commissioned rank, this plan was modified in July 1943. The Commandant felt that from then on there would be sufficient Class V (b) Reservists who "as a result of education, past experience and training can supply the demand and perform the duties as officers. The plan of selecting commissioned personnel, in the main, from the ranks will build up a high standard of morale, efficiency, and esprit de corps." Thus, beginning with the eighth class, in October 1943, the candidates' class was composed of both civilian and enlisted women, with the majority in the latter group. To be eligible, a Marine had to be recommended by her commanding officer. A board of seven members, including the Women's Reserve Director, as well as both regular and reserve male officers, was convened regularly to review and pass judgment on applications from enlisted personnel.(82) This new plan, it should be pointed out, did not completely close the door to civilian candidates. Women with specialized abilities needed by the Marine Corps or those considered to have generally outstanding leadership qualities were still accepted, but on a far more limited basis. The first class of ex-enlisted Marines was graduated on 15 December 1943, and thereafter the majority of new women officers had served in an enlisted capacity before being commissioned.(83)

After the transfer to New River, all officer candidates were appointed to the rank of private first class and remained as such during the entire course, a move that brought the women's training more into line with that of male Marines, since this system was being used in the men's OCS program at Quantico. The cadet category had served its purpose and was dropped once the Marines were training strictly on their own. Meritorious enlisted women who held the ranks of corporal or sergeant temporarily reverted to PFC, and all candidates wore PFC chevrons and OC pins on their uniform lapels and caps. Although the outward appearance of equal rank prevailed, the higher-rated WRs were still eligible to draw the pay of their actual rank. In the event an enlisted woman did not complete the course, she resumed her regular rating. If she so desired, she was eligible after six months to reapply for a new candidates' class "without prejudice against her" because of her earlier failure.(84) At the completion of training, successful candidates were commissioned in the "appropriate" rank. The custom developed of awarding first lieutenant commissions to a small proportion of top candidates and second lieutenant rank to the rest. In several cases, unusually well-qualified candidates were awarded the rank of captain immediately upon completion of candidates' class. But this was no "snap" course; attrition averaged over 30 percent.(85)

Reserve Officer Class

When officer ranks were opened to enlisted personnel in late 1943, it became apparent that even an outstanding NCO did not always make an immediate good personal adjustment to officer status. Then, too, it seemed advisable that these ex-enlisted women should become somewhat more accustomed to their gold bars while still at school and before going out to their first officer jobs.

Accordingly, the first reserve officer class was established after commissioning of the eighth officer candidates' class, in December. Thereafter, the reserve office class was composed of successful graduates of officers' class as well as graduates of the earlier classes who had been on active duty and for whom it served as a refresher course. Principal emphasis was on typical personnel problems. Discussions included realistic problems in administration, recreation, messing, rehabilitation, and the psychology of behavior patterns that a woman officer might have to deal with on the job. The officer training program was thus lengthened to a full three months, with eight weeks of fundamental indoctrination plus the four-week training offered by the reserve officer class.(86)

IV. The Uniform

Design of the uniform for members of the Woman's Reserve had high priority, and the basic ensemble of the uniform was designed prior to actual formation of the Women's Reserve. In mid-December 1942, a memorandum from the Commandant to the Assistant Secretary of the Navy officially requested that Mrs. Anne Adams Lentz, then an employee of the War Department, be assigned to duty at Headquarters "for a period of approximately 30 days."(87) Mrs. Lentz had been employed by the school uniform section of a large New York department store and for eight months had helped the WAACs in the design of their uniform. She came on duty immediately, early in January, and, following consultation with the Depot Quartermaster in Philadelphia, was detailed to New York City to oversee the construction of model uniforms for the Women's Reserve by the Women's Garment Manufacturers of New York.(88) The original concept for design of the uniform was clearly spelled out. The men wanted the women to be dressed in the traditional Marine forest green and to look as much as possible like Marines. Later that month, her original 30-day duty about to expire, Mrs. Lentz decided to stay on. She was sworn in as a WR captain on 18 January 1943, the oath of office being administered by her husband, Brigadier General John M. Lentz, who was attached to Army Ground Forces Headquarters in Washington, D. C.(89)

After Captain Lentz conferred with clothing designers in New York and Marine Corps Headquarters, as well as the supply division in Philadelphia, a general type of uniform was adopted based on tradition, theory, and drawings. Samples were made up of various uniform designs, these were shown to the Commandant and others at Headquarters, and both the winter and summer styles were adopted.(90)

Once the details of design and construction were officially approved, the uniforms were manufactured for the Marine Corps and sold by civilian suppliers under the same general setup as prevailed with the WAVES. Although newspaper clippings showed the new uniform the week after public announcement, and a few key women officers who were constantly in the public eye were issued uniforms almost at once, the inexorable law of supply and demand made it impossible to provide uniforms in quantity until April. Nearly all the early USMCWR officers went on active duty or training, to all outward appearance civilians.

Public interest in the uniform, as well as the name and all other matters concerning the Women's Reserve, was keen. Perhaps mindful of the blue and red of the Marine dress uniform, one woman sent in to Marine Corps Headquarters a picture of a blue and red suit ensemble appearing in the current issue of a high-styled fashion magazine, recommending its adoption and commenting that the hat would "probably have to be modified" for the Marine Corps' purposes.(91) Another unsolicited offer of help came from an ex-Marine of World War I who volunteered his old dress blue cape and wrote: "I am wondering if the uniform of the new Women's Reserve of the Marine Corps will include the blue cape.... I happen to have found two of these among my effects a few days ago....I will be glad to donate them to the new organization if you can use them."(92) Customarily, Marine Corps Headquarters sent individual replies to such letters, thanking the writers for their interest in the Marine Corps and offer of help, but advising that a forest green uniform had already been decided upon for the women's use.

Essentially the regulation Marine uniform was adopted. Its feminine counterpart was identical in color with that of the men, but of a slightly lighter-weight serge or covert fabric, although enthusiasts at the time not infrequently proclaimed that the girls' uniforms were "cut from the identical forest green cloth as the men's."(93)

In accordance with provisions of the law which had authorized the Women's Reserve, a uniform allowance and gratuity of $250 was made for officers and $200 for enlisted. It was expected that uniforms would be purchased, fitted, and paid for during the period of indoctrination. From her uniform allowance, a Woman Marine purchased two winter uniforms, hats shoes, summer outfits, a handbag, a wool-lined raincoat (at $41 the most expensive item in her wardrobe), and various other articles.(94)

Official Issue

The winter uniform consisted of a forest-green, tailored suit with a semifitted, unbelted, three-button jacket with roll collar and notched lapels, worn with a plain matching six-gored skirt that extended approximately to the bottom of the knee cap. The jacket had four pockets, and the traditional Marine pointed-overlay cuff detail finished the sleeve. Dull-finished bronze Marine ornaments were worn on the collar, and jacket buttons were of the same design and finish. A khaki shirt and tie, cordovan oxfords or pumps, seamed beige hose, dark-brown gloves, and a dark-brown shoulder bag were worn. The visored, bell-crowned cap had a large dull-bronze finished Marine Corps ornament in front and was trimmed with a scarlet cap cord. This cord was a striking difference of the women's uniform and replaced the brown chin strap of the men's dress cap. A matching scarlet wool muffler was worn with the trenchcoat or overcoat. Officers wore their rank insignia on the shoulder straps of the jacket and on the shirt collar. They also had the option of white shirts and dark green ties for dress wear. Enlisted personnel wore their chevrons in the manner of male Marines. In most other respects, the uniforms worn by officers and enlisted women were quite similar. (95)

Rather than the traditional military khaki with close-fitting collar and necktie worn by male Marines as well as the women in World War I, the summer work uniform designed for the women in World War II was a tailored two-piece dress, initially of green and white striped plisse crepe, and shortly thereafter of similarly striped seersucker. On the principle that "coolness makes for efficiency," it had an open V-neck, short sleeves, and four patch pockets. Captain Lentz was the one who first suggested breaking with the tradition of summer khaki and the use of seersucker uniforms because of the ease with which they could be laundered, since it was recognized that at many places members of the Women's Reserve would be entirely dependent on themselves for the proper laundering and smart appearance of their uniforms.(96) The single-breasted jacket of this dress had five large white buttons down the front; small white buttons closed the pointed flaps of the four pockets. Commissioned rank was indicated by metal insignia on the shoulder straps and noncommissioned rank by green chevrons. Officers at first wore their insignia right on the straps, but it was soon realized that on the striped material the insignia was not easily seen. Therefore, a slightly stiffened solid green shoulder board in the shape of the strap was devised. Dull-finished bronze Marine Corps ornaments were worn on the collar. At first, the summer headgear was a round cap with a snap brim, but this was soon replaced by a cap of the same style as that of the winter uniform, but of a light spruce-green cotton twill. This cap had a white cap cord and the same large globe-and-anchor ornament. Later, a light-green, garrison-style cap with white piping was authorized. The same dark-brown oxfords or pumps used with the winter service were worn. For summer wear the handbag had a matching spruce-green cover, easily removable to launder. Gloves were white. All items of the summer uniform were designed so they would be washable and easy to keep in good repair.(97)

Summer dress uniform was a two-piece sparkling white cotton, in the same styling as the seersucker uniform, but worn with gold buttons and insignia, white pumps and gloves, and the same green-visored cap and matching cover for the handbag.(98)

Uniform regulations were issued and modified as required. (99) The proper lipstick hue was prescribed as a clear red, or close to the trade shade known as "Montezuma Red" which matched the winter cap cord and muffler and was "neatly and thinly applied."(100) Girdles were a <u>must</u>, no matter how trim or willowy the figure. Slips were to be worn and were not to show below the skirt. Hair might touch but not cover the collar. Finger-nail polish was an option, but if worn had to match the lip-stick. And, unlike their more casual civilian sisters, hats and gloves were required at all times when outdoors.(101)

In addition to regular summer and winter uniforms, certain specialized types of uniforms were issued such as bibbed over-alls for work clothes. An attempt was made to keep all uniforms attractively-styled but simple, so as to keep within the monetary uniform allowance and make them easy to keep cleaned, pressed, and sharp looking.(102)

Those Dress Whites!

Summer dress uniforms easily won hands-down honors from both Marines and non-Marines as the most attractive and feminine uniform of any women's service. They were an immediate hit. When the Third War Loan Drive got underway on 9 September 1943, Women Marines attired in the dress whites nearly stole the show in Philadelphia. "The snappy-looking members of the MCWR--a score of them--were part of the official escort for the dozen or more Hollywood stars," wrote one observer. "All the stars high-ly complimented the uniforms of the Women's Reserve...Dick Powell said he thought they were the nicest he had seen." Perhaps the best accolade of all came from the policeman who commented simply: "I hear from all sides that the Women Marines outshone the stars."(103)

Special Uniform Class and Uniform Distribution

From September through December 1943, 13 women officers were attached to Headquarters for intensive training in the various phases of tailoring, alternations, clothing construction, and fitting. Upon completion of a six-week course, they were assigned to uniform shops being operated by Post Exchanges at major Marine Corps posts throughout the country. Two phases of training were covered in the course: materials, design, con-struction, specifications, and uniform regulations; and administration and successful operation of a uniform shop, in-cluding the set-up of a Post Exchange stock control system. Unlike some of their sister services, the Women Marines' clothing was not government issue. Regulation clothing and all items of uniform were purchased by the Post Exchange, and in turn bought by the women, using the allowance given them by the government. At the Post Exchange Uniform Shop, Women Marines especially trained for the job fitted the clothing.

Originally, in 1943, Marine Corps Women's Reserve uniforms had been manufactured by various firms and sold to retail outlets. These stores then sold the uniforms to the individual women, a system of supply and distribution which had been used by the WAVES and adopted by the Women Marines. However, since it seemed to have a number of inherent difficulties, including chronic shortage of popular sizes, a change was made on 16 February 1944, when responsiblity for distribution of the women's clothing was placed in the Quartermaster Department which supplied the Post Exchange shops throughout the country for the remainder of the war.(104)

Uniform Board and Regulations

On 11 June 1943, a Uniform Unit was established as part of the Women's Reserve Section at Marine Corps Headquarters. Its purpose was to provide for the complete uniforming of the individual at the time of assignment to active duty. A Uniform Board which suggested articles of clothing and made recommendations to the Commandant was established on 17 June. A complete list of uniform regulations, including explanatory sketches, was issued in July 1943, after having been approved by the Uniform Board, the Commandant, and the Secretary of the Navy. These regulations were later modified and reissued in April 1945.(105)

On 16 June 1944, the Uniform Unit of the Women's Reserve Section was transferred to the Supply Division, Quartermaster Department. A number of steps were taken to make the entire system of supply and distribution of uniforms more expeditious. In October, this division took over the writing and approval of all specifications for Women's Reserve clothing. This job had previously been done by the Philadelphia Depot of Supplies which had based many of its supply projections on its past experience of working with men's clothing, a system which not too surprisingly proved inadequate, as certain characteristics of women's clothing were entirely unrelated to men's.(106) Although the Woman Marine uniform itself was well-accepted and a definite success, many of the administrative procedures concerning its design, specifications, accurate sizing, inspection, and distribution remained a changing but constant problem. As the Director herself once commented: "...the supply of MCWR clothing was one of the few problems to which a satisfactory solution had not been found at the time that demobilization began."(107)

V. Jobs and Job Assignments

All Marine Corps training for the women, whether basic or specialist, was tough and thorough. The objective was to indoctrinate a Woman Marine so completely in her field or specialty that she could handle any contingency that might develop in the day-to-day job situation. Skill and precision were mandatory, whether in transcribing shorthand notes or packing parachutes. As with the battlefield Marines, the women learned that mistakes could cost lives.

Women Marines in World War II had many advantages over their predecessors in World War I. One of the biggest of these was in job assignments and increased job responsibility. The women Marines of 1918 (or, as they were called in the parlance of the day, "Marinettes") (108) numbered 305 and had primarily clerical duties--stenography, typing, bookkeeping, and messenger chores.(109)

By contrast a much wider range of jobs were available to USMCR women in the Second World War. This included such diverse tasks as being a radio operator, photographer, parachute rigger, motor transport driver, aerial gunnery instructor, cook and baker, quartermaster, Link trainer instructor, control tower operator, motion picture technician, automotive mechanic, teletype operator, cryptographer, laundry manager, and post exchange manager. In addition, as the women's units were set up in posts thoughout the country, there were hundreds of "line" or company work assignments created which compare roughly to personnel management jobs in civilian life, as well as the inevitable stenographic and related desk jobs.

Early recruiting literature in 1943 had referred to "more than 30 different job assignments."(110) Actually, this turned out to be an extremely modest estimate. Once on the job, the women proved themselves so versatile that they were soon performing assignments previously considered strictly in the masculine domain. They also took on, informally, other duties such as swimming pool lifeguard on some posts and stations.(111) The total number of different job classifications turned out to be more than 200. (See Appendix A for complete list.)

Job Classification

A job classification system was established in March 1943 so that each Woman Marine could be easily placed in the task she was suited to handle following completion of her indoctrination course. At both the U. S. Naval Training School in the Bronx and the U. S. Naval Reserve Midshipmen's School in Northhampton, women Reservists were questioned as to professional experience, education, hobbies, and linguistic ability. Tests determined their special aptitudes.(112)

The classification system, paralleling that used for the men, was under the direction of Captain Cornelia Williams, at Marine Corps Headquarters. She held a Ph.D. degree in psychology from the University of Minnesota and had wide experience as a college instructor and administrator in student personnel work.(113) The Women's Reserve Section of the Detail Branch was responsible for the classification and detail of all Women Reservists. The section's classification work involved selection of tests, designing a qualification card, supervising the selection and training of classification personnel, and analyzing jobs and giving them appropriate specification serial numbers. The Women's Reserve Section projected and planned the distribution of all Women Marines in accordance with the current needs of the service and requested the necessary orders to send them to the appropriate schools or duty stations. This required analysis of billets, an analysis of available personnel, and the matching of the two as well as possible.(114)

After the transfer of recruit training to Camp Lejeune in early July, each new Marine was tested and interviewed during her first week of training. Assignment to jobs and eligibility for the specialists schools was determined by the classification section at the Women's Reserve Schools. Eventually, classification specialists were assigned to all posts and stations to assist in the assignment of women who reported for duty and to reclassify them when necessary. Reassignment sometimes became necessary because the personnel were misassigned in the first place or because the needs of the service had changed.(115)

More than half of all Women Marines were assigned to office jobs where they utilized their civilian experience. A statistical breakdown of the 17,672 women on duty at the end of the war shows they had primary military job specialties in the following categories:

Clerical and sales	11,020	(or 62.4 percent)
General duty	1,648	(9.3 percent)
Mechanical	1,371	(7.7 percent)
Professional and managerial	1,342	(7.6 percent)
Semi-skilled jobs	1,305	(7.4 percent)
Agriculture and service	587	(3.3 percent)
Student	35	(0.2 percent)
Unskilled	14	(0.08 percent) (116)

An analysis of military assignment in relation to civilian background shows that:

(1) The total number of Women Marines assigned to clerical duties was about the same as the percentage so employed in civilian life.

(2) Proportionately fewer women were utilized in "professional" categories in the Marine Corps than came from professional jobs in civilian life. Most of the discrepancies can be accounted for by the relatively large number of civilian school teachers who enlisted in the Women's Reserve, in contrast to the number of instructional jobs available.

(3) Proportionately more women were used in the Marine Corps in mechanical jobs than came from these types of jobs as civilians --especially in aviation.

(4) More women were used in the service category than came from this category as civilians. Here again the discrepancy resulted from the fact that all commissary jobs in the Marine Corps were classified as personal service.(117)

Promotion

Promotion is always a difficult problem, and enlisted promotion in the Women's Reserve was no exception. Several different plans were tried, found not entirely satisfactory, and were amended. Eleven Letters of Instruction were issued on the subject in two and a half years. The final system used for line personnel which seemed to work best provided that 75 percent of the combined strength of privates and privates first class could rank as privates first class. A quota of promotions to the fourth and fifth pay grades was alloted monthly to each post, and these promotions were made by the post commander after tests were given and successful candidates determined. Promotions to the first three pay grades were recommended by commanding officers and effected by Marine Corps Headquarters as vacancies existed. In the specialty schools top-ranking students were often graduated a full rating ahead of the rest of the class.(118)

With officers, seniority was the chief determining factor in the beginning. Later, it was decided that promotion should be made by selection, and for specific billets, so it would be possible to have the rank where it was most needed. The principle of bloc promotion from second to first lieutenant was adopted and all Women's Reserve officers who served satisfactorily in commissioned rank for at least 18 months were finally assured one promotion.

"Spot" promotions were authorized in cases where a woman with specialized skills was needed to fill a billet which by table of organization called for a higher-ranking Marine. Such promotions became official only when the woman later came up for regular promotion with contemporaries from her same training class or like seniority, and was selected for promotion to the next higher rank.(119)

The Buildup

In May of 1943, as the first Hunter and Mount Holyoke classes were graduated, Women Marines began reporting in small units to camps and bases, as well as to Headquarters in Washington, where there were approximately 250 enlisted women and 15 officers by mid-May.

A little later, as the Marine raiders headlined the front pages with their landing on New Georgia and the capture of Viru Harbor, the first two Women Marines assigned to motor transport completed their specialist courses at Camp Lejeune and climbed into military trucks.

July--as the new WR Schools complex began in full schedule at Camp Lejeune--found the Women Marines there moving full force into Paymaster and Quartermaster schools, well-aware that 9,000 miles away the Central Solomons were being blasted by Marine fliers and that more men would be needed for the island-hopping conquest of the enemy.

August found the Women Marines at Lakehurst, New Jersey, learning to fold the silken safety of parachutes--and for them, the vital importance of the rigger's work was underlined by the landing of Marine Corsairs on Munda Airfield, a promise of air-battles to come.

September found three battalions of Women Marines training simultaneously at Camp Lejeune's Recruit Depot. As the first Marine planes landed in September on the newly-constructed air-field at Barakoma, Vella Lavella Island, in the Solomons, 25 women skilled in the handling of the Link trainer began instructing future Marine pilots at the air station at Edenton, North Carolina.(120)

By the following month, the first Women Marines had reported for duty on the West Coast--at Camp Pendleton and the air stations at Santa Barbara and El Centro, California. At Cherry Point, where the women were already established, the entire bus system was taken over by them. With the buses, they inherited the responsibility of dispatching, maintenance, and repairs. And at New River's "Tent City," combat Marines were surprised one morning when a group of Women Marines electricians reported to wire an area of Dallas huts.(121)

In November at San Diego and Parris Island, male teletype operators, cashiers, stenographers, and file clerks left in large numbers for the front lines as the women reported for duty. At the Marine Corps Air Station at El Toro, California, the first 97 women paved the way for the thousands who eventually reported there to work in offices, handle mail, drive jeeps, repair aircraft engines, and instruct combat crewmen in aerial gunnery. At the same time, at Quantico, Virginia, the first four women radio operators reported for duty.(122)

Reported the Quantico *Sentry* in November 1943: "The women Marines have landed. Quantico--the beehive of training in World War I, on Wednesday received the first of the WR detachment. Over 11,000 green-clad Women Marines (approximately 8,500 already on active duty) are training and working at 125 different types of jobs at 52 other Marine posts and stations."(123)

Other Women Marines, in training at the University of Wisconsin, at Miami University in Oxford, Ohio and at Omaha, Nebraska were learning the International Morse Code and the maintenance and operation of sending and receiving equipment, in preparation for the eventual take over of station-to-station communication at posts throughout the country.(124)

Shortly thereafter, in December 1943, the significance of air power was re-emphasized to the women in forest green by the first Marine fighter sweep on Rabaul. With approximately one-third of their total strength destined for some phase of aviation, Women Marines later that month moved into the fields of celestial navigation, studying mathematics and theory at Hollywood, Florida and the operation and maintenance of the miniature bomber at Quonset, Rhode Island. At the same time, control tower operators, trained at the Atlanta, Georgia specialty school, were assigned to Marine flying fields, releasing more men for the air units in the Pacific.(125)

Thus, by the end of their first year in service--13 February 1944--the Women Marines were nearing their planned enlistment strength and were a close-working unit on each camp, base, or air station. In quartermaster departments, they were taking over not only the paperwork, but the actual loading duty in stockrooms. In other activities, they were aerial photographers and darkroom technicians, welders and painters, telephone operators, and aircraft and instrument mechanics.(126)

Watching the combat men ship out for battle zones, the Women Marines worked steadily at their jobs and assumed new ones. June and July of 1944 marked the battles for Saipan, Tinian, and Guam--and the casualty lists passed through the hands of the Women Marines stationed at Marine Corps Headquarters. Billeted at Henderson Hall in Arlington, Virginia, they were part of the over 2,000 on duty at Headquarters as stenographers, typists, clerks, and messengers. Working in procurement, aviation, mail and files, plans and policies, and other offices, they handled the clerical details of muster rolls, decorations, statistics, payrolls, identifications, and other activities attendant upon the tremendous task of administration.(127)

In September, the landings on Peleliu once more brought home to the women the need for constant replacements and supplies in the Pacific. Cherry Point graduated its first class of women skilled in PBJ (Mitchell bomber) repair, and women radio operators began standing watch at the field's lighthouse tower. At the same time, WR aviation machinists were graduating from the Naval Air Technical Training School in Norman, Oklahoma, and reporting for crew work at airfields. In the supply depots in San Francisco and Philadelphia, women packed and repaired radio parts, sorted clothing, and drove trucks.(128)

Beginning in December and though the first half of 1945, the participation of the Women Marines in the great push to victory was made even more complete when the passage of modified regulations permitted them to serve overseas. Nearly a thousand Women Marines served in Hawaii, at the Pearl Harbor Naval Base and Marine Corps Air Station in Ewa. Here they did much as they had done in the States--moved into offices, workshops, and other installations, freeing combat men for front-line duty or return Stateside for well-earned furloughs. With the ending of hostilities and surrender of Japan, nearly 20,000 Women Marines in jobs both in Hawaii and the States knew that their contribution had indeed been a vital one. (129)

Job in Aviation

Before the war ended, nearly one-third of the Women Marines had served in aviation at Marine air commands and bases. Under the special arrangement that the Division of Aviation had within the Marine Corps whereby it trained, assigned, and supervised its own personnel, this same policy was extended to women. Upon completion of recruit training, they took classification tests and were divided into two main groups: those assigned to aviation and those to non-aviation, or general duty. In most cases the Division of Aviation then made its own arrangements about specialty training.(130)

All personnel working in aviation--whether in the "glamour" technical assignments such as Link trainer instructors and control tower operators or in the purely administrative functions such as stenographers and stock clerks--were classified as holding aviation jobs. Since so many aviation jobs were being filled by women, it became essential to have at least one key officer responsible for the varied liaison and training duties. A memorandum in early 1943 cited the modest requirements needed by this woman: she should have, preferably, both aviation and general business experience plus the executive ability to work with the Division of Aviation in connection with liaison, organization, procurement, and training. Also, since Marine aviation traditionally is so closely linked with naval aviation, an understanding of Navy Department organization, as well as that of the Marine Corps, was also considered desirable. A main requirement was described simply as the ability to "handle problems and get things done."(131) Two women were selected for special duties in the Division of Aviation and approved for appointment to the rank of captain, following completion of officers' training. They were Marion B. Dryden and Katherine D. Lynch, both members of the fifth officers' class, who were commissioned on 20 September 1943.(132)

Indication of how the women replaced the men in air ground jobs can be seen from the record at Cherry Point. By August 1944 all the training in Link instruction was handled by the Women Marines. They took almost complete charge of the photography department and film library. Ninety percent of the parachute packing, inspecting, and repairing was done by the women, and 80 percent of the landing-field control tower operations were being "manned" by women.(133)

With assignment of large numbers of women to duty at Marine air stations, a number of detachments were activated as Aviation Women's Reserve Squadrons. The function of these units was to supply various technical and administrative personnel needed by the male Marine operational training unit of the next higher echelon. Aviation Women's Reserve Squadrons in operation at the end of the war included: Number 1 at Mojave; Number 2 at Santa Barbara; Number 3 at El Centro; Numbers 4 and 5 at Miramar; Numbers 6-10 at El Toro; Number 11 at Parris Island; Number 14 at Ewa, Hawaii; Numbers 15-20 at Cherry Point; and Number 21 at Quantico.(134)

"Appropriateness" of Jobs

A four-fold classification of the "appropriateness" of jobs in respect to innate female capabilities to perform the work in contrast to the men they released was made, based on World War II evidence. These are the following classifications and the conclusions which were reached.(135)

Class I: Jobs in which women are better, more efficient than men. Example: All clerical jobs, especially those involving typing or requiring fairly routine tasks but coupled with a high degree of accuracy in the work; administrative jobs connected with organization and administration of the Women's Reserve; and instructional jobs of all types.

Class II: Jobs in which women are as good as men, and replaced men on a one-to-one basis. Examples: some clerical jobs in which men are especially good, such as accounting;

some relatively unskilled service or clerical jobs, such as messengers or Post Exchange clerks; some of the mechanical and skilled jobs, such as watch repairman, fire control instrument repairman, tailor, sewing machine operator--especially those jobs requiring a high degree of finger dexterity.

Class III: Jobs in which women are not as good as men, but can be used effectively when need is great, such as wartime. Example: most of the jobs in motor transport--men are better as motor mechanics and even as drivers when the equipment is heavy and the job demands loading and unloading as well as driving, as it often does; most of the "mechanical" and "skilled" jobs; supervisory and administrative jobs, such as first-sergeant (except in WR units) where maximum proficiency depends on years of experience in the Marine Corps, and also some supervisory jobs where part of the personnel being supervised is male; strenuous and physically tiring jobs, such as mess duty where experience showed that more women had to be assigned to cover the same amount of work because they could not endure the long hours and physical strain without relief as well as men.

Class IV: Jobs in which women cannot or should not be used at all. Example: jobs demanding excessive physical strength, such as driving extremely heavy equipment, stock handling in warehouses, heavy lifting in mess halls; jobs totally inappropriate, such as battle duty or jobs requiring that personnel be engaged at particularly unfavorable hours, jobs protected by special civil service regulations for civilians, such as librarians.

The Philosophy of Hard Work

The fundamental purpose of the Women's Reserve in World War II was to train Women Marines to replace men in essential duties at Marine bases, without loss of military efficiency. The Marine Corps had no place for self-appointed glamour girls. Enlistees were told bluntly they must be ready to learn many new things and to put up with a lot of hard work. Typical of this straight-forward, realistic approach was a statement made by the Director less than six months after the Women Marines had been aboard. In an official memorandum she noted that women of the Marine Corps were now stationed all over the United States, serving in all kinds of jobs, and true to traditions of the Corps, "cheerfully assume whatever duty may be assigned to them, even though it may be a job they do not particularly like. Our people as a whole do just as the other members of the Marine Corps and take satisfaction in a hard job well done."(136)

Admittedly some of the men had second thoughts about the real usefulness of the women until they saw them in action and observed "dungareed WRs tear down a Corsair engine, or slide out, greased and grimed, from under a six-by-six truck, or handle a fouled-up traffic pattern from a control tower with the same ease as they did a typewriter."(137)

Not unexpectedly, morale was highest among those women who could see that they had actually "freed a man to fight" or that their efforts were a direct help to fighting Marines. Morale was correspondingly low where the women were not kept sufficiently busy or where their jobs bore no visible relation to the war effort--such as beauty operators. The greatest single morale factor among members of the Women's Reserve was job satisfaction. (138)

And at the end, men who weren't too enthusiastic about admitting women to "their" Marine Corps were not at all enthusiastic about seeing them go. As one captain ruefully remarked: "You can't pick good clerks out of thin air. The women have done remarkably well."(139)

VI. Administration and Policies

The original Plans and Policies study which had recommended the formation of a Women's Reserve also suggested that it be placed for administrative purposes in the Division of Reserve of the Adjutant and Inspector's Department. This was a logical decision since the Division of Reserve was responsible for the procurement of all Marine Corps reserve personnel. A newly-created unit, called the Women's Reserve Section, was attached to the Division of Reserve to handle matters dealing with administration of the Women's Reserve, *e.g.*, training, uniforming, and regulations. Suitable personnel were placed in the Women's Reserve Section to handle the new activity. In addition, a senior woman officer was assigned to major activities at Marine Corps Headquarters which handled matters affecting women--such as Personnel, Administrative Division, Public Information, Plans and Policies, and Supply. (140)

It was believed that women could be most useful to the Marine Corps if they were regarded for purposes of organization much like "extra" Marines. Thus, all administrative action relating to them was taken through the regularly established divisions which were already performing such functions for the men, and the Women's Reserve was never organized as a separate administrative unit. (141)

Initially, the Director, MCWR, was charged with the "...procurement, instruction, training, discipline, organization, administration and mobilization of the Women's Reserve for the duration of the war and six months thereafter."(142) From 13 February to 29 October 1943, she was attached to the Women's Reserve Section for "purposes of instruction" as she learned her way around military procedures. On the latter date she was transferred as a Special Assistant to the Director of Personnel, with her chief duty being that of advising him on policy matters concerning the Women's Reserve. Although the Director had considerable influence in developing policies and procedures for the new Women's Reserve, actually she never took any independent action regarding the administrative handling of the Reserve. She made the recommendations to the Director of Personnel who, in turn, was authorized to take appropriate action.(143)

Since the guiding philosophy was to treat the Women Marines, for administrative purposes, much like additional Marines, it was logical that those regulations governing the men which were appropriate and practical for the women would also be adopted. Some administrative procedures and policies were also adopted from the WAVES. This, too, was natural, since in the beginning recruiting and training of the Women Marines had been conducted in conjunction with that of the WAVES. Then, too, the Marine Corps Women's Reserve benefited from the valuable experience of the other women's services which was freely shared. The Canadian Women's Army Corps was also most helpful. In fact, an officer of the regular Marine Corps, Lieutenant Colonel John B. Hill, had paid an official visit to the CWAC and the other Canadian women's services in January 1943, before the Marine Corps Women's Reserve was formed, to learn first-hand about the curricula, personnel policies, and other organizational details that might be helpful to the new American service.(144)

Cooperation with the Women's Services

Despite the fact that the women's services were competitive, in the sense they were all eager to enlist well-qualified candidates, a high degree of cooperation and good will existed between the directors themselves. The women leaders of the three other services, and their highest ranks, were: Colonel Oveta Culp Hobby, WAAC and WAC; Captain Mildred H. McAfee, WAVES; and Captain Dorothy C. Stratton, SPARS.

A typical matter on which the four women's service leaders worked together was drawing up a unified program of the recruiting, as well as enlistment, of women from the war industries, civil service, or agriculture. After the four directors had worked out an agreement which resolved their own differences of viewpoint or emphasis, the recommendation was then submitted to the Joint Army-Navy Personnel Board for final approval. The board, in turn, issued the all-service policy which was then followed by the four women's reserves.(145)

As a general policy, the enlistment of applicants already employed in any of the war industries was discouraged. The case was referred to the local office of the United States Employment Service which had to authorize a release. In instances of civil service workers who sought enlistment in the Marine Reserve, the policy adopted was that the woman had first to secure a written release from the agency. An employee who was released "without prejudice" on the part of her employer could apply to the Women's Reserve under the same conditions as a non-Civil Service employee. On the other hand, an employee whose resignation had been accepted "with prejudice" and whose employer was reluctant to have her go, was ineligible for membership in the armed services until 90 days had expired from the date of acceptance of her resignation. Civil service employees who resigned to enlist in the Marines were not returned to duty at their former place of employment, even if they happened to be classified with a military job description identical to their previous civilian occupation.(146)

Typical of the good feeling that existed among the women's services was the Marine Corps' three-day Open House at Camp Lejeune held from 13-15 October 1943, after transfer there of all the women's training activities. Planned as a method of information exchange between the women's services, the three-day event included inspection of training facilities and methods as well as observation of the performance of Women Marines on the job. Representatives of the WACs, WAVES, and SPARS all attended as well as many high-ranking male officers of the corresponding services.(147)

Policy about Assignment and Housing

From the beginning, the Marine Corps decided that Women Marines would be assigned "only to posts where their services have been requested."(148) The matter of proper housing facilities, in connection with assignment to duty, was a major consideration. The early November 1942 Plans and Policies survey, which had sought estimates on the number of women needed by different posts had also requested information regarding the quarters available for their use. Women were not to be assigned to posts lacking proper housing "unless they can be quartered with the WAVES or other satisfactory arrangements can be made." (149)

It was also a policy that no less than two women would be assigned to a station or sub-station, a move designed to "prevent loneliness and obviate possible unfavorable comment."(150) No enlisted women were to be assigned to a post unless a woman officer was present or in the near vicinity. As a matter of practicality, it became the general rule not to assign a Woman Marine officer to units of fewer than 25 women. The obvious exception, of course, was in assignment to procurement offices in large cities. The officer-to-enlisted ratio was projected at 5.7 percent.(151)

Upon completion of their training, women were assigned to duty on posts and stations where they were under the authority of the commanding officer of their unit, who in turn reported to the commanding officer of the post. In respect to their quarters, mess facilities, and general administration the women Marines were usually a relatively autonomous unit. Women Marines living on the regular posts had their own barracks area which they maintained themselves. In cases where MCWR personnel were stationed in cities, the question arose whether to obtain barracks for them or put them on subsistence, a monetary allowance to compensate for food and living costs. Where only a few were on duty, such as the procurement stations, the women were naturally put on subsistence. In Washington, D. C., where as many as 2,400 Women Marines were on duty, Henderson Hall was build and operated as an independent post. The health, feeding, military attitude, and discipline of the Women Reservists were all improved when barracks were available for them. This was due to the fact that when living together as a military unit, the women felt more like Marines than they did when they lived scattered throughout a duty area, and they also enjoyed more of a sense of comradeship with one another.(152)

In many cases, the work of the women was supervised by male officers, but every-day matters of discipline and command questions were left to the Women's Reserve officer. Despite the unique theoretical concept of dual supervision, in practice it usually worked well. In the few instances of serious disciplinary problems, male commanding officers usually sought the advice of the senior Women's Reserve officer on their posts before handling the matter.(153)

Assistants for the Women's Reserve

In the fall of 1943, the buildup of the Women's Reserve witnessed the assignment of thousands of women to far-flung posts throughout the country. It became imperative to have some type of regular reporting system so that Marine Corps Headquarters would know immediately of all pertinent matters beyond those which were vital enough to be committed to official correspondence, those learned first-hand by the Director's field visits, or those heard informally through the military's oldest information media, the grapevine.

Accordingly, the senior woman officer at stations where Women Marines were serving was designated as an "Assistant for the Women's Reserve."(154) She was responsible for keeping in close touch with the Director and advising her on all matters of welfare, health, jobs, training, housing, recreation, and discipline. The Assistant for Women's Reserve activities was likewise responsible for keeping the post commanding officer informed on anything that pertained to the women under his jurisdiction. The procedure of a monthly written report was instituted. This was sent every month by the post Assistant for the Women's Reserve to Marine Corps Headquarters, with a copy to the post commanding officer. It contained information on all aspects of the women's jobs and well-being as well as "full remarks concerning items of special interest at the station."(125) These monthly reports supplemented personal visits of the Director, who made it a practice to spend a quarter of her time away from her Washington desk to see for herself how the various units of the Women's Reserve were operating throughout the country.(156)

Policy about Women's Authority

The authority of the women officers was exercised "over women of the Reserve only" and was "limited to the administration of the Women's Reserve."(157) In everyday practice it had been determined that the "relationship of women officers or noncommissioned officers to enlisted men in the administration of their work is similar to that of a civilian teacher in a military school. While the woman officer may give instructions in connection with the work, matters of discipline should be referred to the man's commanding officer."(158)

Although the phrase "matters of discipline" was also interpreted informally as "matters of job performance," the proper scope of the authority of women officers when they were assigned to duties involving supervision over male personnel continued to cause some uncertainty. Some months later the Commandant felt it necessary to issue further clarification.

"It appears that the services of officers and non-commissioned officers of the Marine Corps Women's Reserve are not being utilized to the fullest extent due to some doubt as to the scope of their authority," he observed in March 1944. This matter has been considered by the Navy Department, he continued, and "it is concluded that it is entirely proper for a woman officer to be assigned to duty subordinate to a commanding officer and her directions and orders in the proper performance of such duty are the acts of the officer in command, even though such orders are directed to male personnel."(159) Thus, within discretion of individual male commanding officers, the door was opened for Women Marine officers to be detailed to duties such as adjutant, assistant adjutant, personnel officer, or mess officer "where the directions and orders necessary in the performance of such duties" were "considered as emanating from the commanding officer."(160)

Changing Policy on Marriage

Originally, when recruiting opened in February 1943, a Woman Marine could be either single or married, so long as her children were not under 18 and she was not married to a Marine. This regulation had been issued by the Secretary of Navy and applied equally to prospective WAVES, SPARS, and Marines, none of whom could be enlisted or appointed if their husbands were in the same service. Furthermore, a member of the Naval Reserve "could not while in service marry an officer or enlisted man in the same service."(161)

Early in March 1943, however, the Secretary of the Navy approved a modification of these existing rules which allowed a member of the Naval Reserve to marry after she entered the service. Single women merely had to execute an agreement not to marry "during the period of their indoctrination or training."(162) Naturally the question arose as to what was meant precisely by the qualifying phraseology. Here the Marine Corps developed a rather generous attitude, decreeing that a woman's basic training (either boot or candidates' school) constituted her "period of indoctrination or training" rather than her entire period of training which often included an additional month or so at a specialist school. The distinction was made because indoctrination lasted for only six weeks during which the women were "learning the principles of military life" and the fact that obligations assumed thereafter by the women "must be secondary to their obligation to the Marine Corps."(163) On the other hand, specialist schools often lasted as long as four months and it was felt that "at the rate that men in the military service are now being sent overseas, this delay would often mean that the couple could not get married at all."(164)

The policy about marriage was modified again in late 1943 when wives of Marines "below the rank of second lieutenant" were allowed to enlist. Originally, the Marine Corps had not concurred with the WAVES in the preliminary discussion that led to this decision, and thus added the following warning when the change was effected: "Each wife shall be made to understand that the probability of being stationed with her husband is very slight, and that consideration cannot be given to personal desires in the matter."(165) On balance, the record shows the Marine Corps tried to steer a reasonable, realistic course between outright forbidding of service marriages, which might simply aggravate other problems, and being too lenient about widespread marriage which might, in turn, easily work out not to be in the best interests of the service itself.

Discipline and Morale

For the most part, discipline was administered according to the rules already established for the men of the Marine Corps. Women, however, could not be put in brigs or prisons, but were confined to quarters. A Women's Reserve officer in the discipline division reviewed all disciplinary cases and consulted the Director for further recommendations. Within the first year and a half, nearly 90 percent of the women were organized into battalions and squadrons, under command of their own women officers. At this time, the commanding officers were given the authority to convene deck and summary courts martial. Women officers were assigned to courts which tried members of the Women's Reserve. Punishment included confinement to quarters, loss of pay, reduction in rank, extra police duties, and, in extreme cases, dishonorable or bad conduct discharge. Pregnancy was considered a medical rather than disciplinary case.(166)

Much thought and effort were given to trying to maintain morale at a high level so that disciplinary action would not be necessary. Recreation and educational services were considered very important in this respect. The necessity of discipline and high standards in every aspect of behavior and work was stressed from the time a recruit set foot in camp. The Women's Reserve subscribed to the philosophy that a "slack ship" is not a happy ship, let alone an efficient one, and directed its personnel and regulations accordingly. It was found entirely possible to maintain high standards in an organization of women and still be humane and understanding in dealing with them.(167)

Officers were thoroughly indoctrinated with the principle that they must be readily available to their enlisted women for "informal personal counsel and advice on matters other than military."(168) Many officers set aside a regular time, often from 1630 to 1700 daily, or at least several times weekly, so that any woman could get private counsel. This tended to keep morale high and reduce problems of adjustment to military life which otherwise might result in disciplinary troubles. The importance of keeping personnel well-informed was also stressed. The guiding philosophy expressed by Colonel Streeter was that "the most able commanders, be they men or women, are those who take the best care of their people and who keep them out of trouble by anticipating the problems that may confront them."(169)

VII. People in the Program

For decades the Marine Corps has prided itself on the colorful and unusual personalities it seems to attract to its ranks. This situation held true for the women in World War II, many of whom seemed to have a special sense of derring-do and *esprit*.

The slogan "Once a Marine, always a Marine" was true for two women who had served as Marines during World War I and again reentered. One of these was Mrs. Martrese Thek Ferguson, a member of the first candidates' class to train at Mount Holyoke College in May 1943. Not only was Mrs. Ferguson a member of the class, but she led it, graduating in the number one spot and being commissioned a first lieutenant. She could boast too, not only of Victory and Good Conduct medals from World War I but of two service sons, one of whom was a Marine. She later rose to the rank of lieutenant colonel and was commanding officer in charge of more than 2,000 women at Henderson Hall.(170)

Another woman who wore the Marine green uniform in both wars was Mrs. Lillian O'Malley Daly. One of the eight women who came to the Corps directly from civilian life in January and February 1943, she was immediately assigned to Camp Pendleton, California, where she served as the West Coast liaison officer. Captain Daly was stationed there at the time Major Streeter and her public relations assistant, First Lieutenant E. Louise Stewart, made their first nationwide recruiting trip in February and March 1943. It was at Camp Pendleton and nearby Camp Elliott that these three new women officers tried their skill at the rifle range, jumped from the parachute tower, flew with paratroopers making their first jump, and rode in tanks.(171)

Another officer who shared a similar background was Major Helen G. O'Neill, who was also one of the first Women Marines in World War II. She had been a Chief Yeoman in the Navy during World War I. She had the distinction of being a 25-year civilian employee of the Navy Department. Prior to her commissioning in the Women's Reserve she had served as secretary to four Assistant Secretaries of the Navy. She had organized the National Yeoman F, a group of ex-servicewomen of the U. S. Naval Reserve from World War I and had served as one of its top officers. Major O'Neill was also a linguist who had studied French, Spanish, and Latin, and was later to concentrate on Russian.(172)

Captain Frances W. Pepper was another member of the Marine Corps Women's Reserve with an unusual background. Graduating as top-ranking member of the third candidates' class in June 1943, she was appointed as a captain, the first woman to receive this rank from any class of aspiring officers. Her service to the Marine Corps dated from 1923, when she joined the Adjutant and Inspector Department of Headquarters. Her work there dealt with appointments, retirement, discharge, and promotion of all commissioned and warrant officers on active duty in the Marine Corps and Marine Corps Reserve. Captain Pepper held a Bachelor of Law degree, was admitted to the District of Columbia Bar in 1931, and later did post-graduate work in the field of international law. During World War I, she served with the Young Women's Christian Association at General Pershing's Headquarters in France.(173)

As a civilian in World War I, Major Helen N. Crean won a Croix de Guerre for heroism under fire. She organized a canteen for the Fifth Regiment of Marines at Naix-au-Forge and the Verdun sector and, working though the hospitals, secured information on the wounded and missing. Later she served at the Red Cross dressing station at Glorieux, France, when it was badly bombed and machine gunned--and her fearlessness won her the French medal. In World War II, she was commanding officer of the Women's Reserve unit at the Marine Corps Air Station, Santa Barbara, California, and later at the air station in Ewa, Hawaii.(174)

A woman reservist with an international background was Charlotte Day Gower, named director of training for the Women's Reserve. Formerly Dean of Women at Lingnan University, in Hong Kong, China, she had been there when the Japanese assault began in that city and had organized first aid stations and helped in rescue work. A prisoner in a Japanese internment camp for five months, Major Gower had taught Chinese to fellow inmates and was later repatriated in an exchange of prisoners.

Trapped in a similar set of circumstances was Staff Sergeant Mary Virginia Herst, of Argonia, Kansas. She was a home economics teacher in Bangkok, Thailand when the Japanese attacked and spent nine months in a prison camp before being repatriated. She was later attached to the Marine Corps Air Station, El Toro, California.

The Pacific Theather of Operations was also more than a headline to Marine Private First Class Peggy Urzendowski. Having spent most of her life in Singapore, Hong Kong, and the Philippines, she was on Corregidor during two fierce bombing raids before being evacuated on a troop ship enroute to New Zealand. From there she was sent to Australia, then Hawaii, and finally the United States--where she quickly enlisted in the Women's Reserve and was assigned to duty at the Marine Corps Base, San Diego, California.(175)

The Women's Reserve Band

If the Women's Reserve listed some unusual personalities in its midst, an unusual and famed institution that came into being during the war years was the Marine Corps Women's Reserve Band. Organized at Camp Lejeune with an initial allowance of 43 women, it was formed with the high-spirited goal of becoming "the most outstanding female band of the country." (176)

To secure the best possible musical talent, the Marine Corps wrote letters to more than a dozen prominent and well-established music schools and colleges, acquainting them with the band and asking them to recommend possible candidates. As a result many women applied for the band before enlistment and joined the service with the express purpose of becoming a member of the band. Women already enrolled in the Reserve also had an opportunity to try out for the band and, if sufficiently talented, to be accepted. All prospects were screened and severely auditioned before being selected as members. The band was organized in November 1943 by Captain William F. H. Santelmann, and was trained by musicians of the U. S. Marine Band. Its director was Master Sergeant Charlotte Plummer, who prior to her enlistment had been director of music in the Portland, Oregon public school system and a member of the city's municipal band.(177)

The band's home base was Camp Lejeune, where it played at the weekly Saturday morning MCWR Recruit Depot reviews. It also performed on every Marine Base on the east coast and toured many cities, playing before such distinguished personalities as President Franklin D. Roosevelt and Admiral Chester W. Nimitz. At the request of the U. S. Treasury Department, the band took part in two war bond and victory loan drives. Its theme song "March of the Women Marines" was written especially for the band by Master Sergeant Louis Saverino and Technical Sergeant Emil Grasser. Probably the red-letter day for the band itself was in October 1945 when, on tour in Washington to take part in the Nimitz Day parade, it serenaded the Commandant, General Alexander A. Vandegrift, outside his office in the Navy Annex.(178)

Quantico's Drill Team

One military accomplishment long connected with the Marine Corps in the public's mind has been excellence of drill and military bearing. While all Women Marines took pride in their ability to maintain this reputation, one group particularly distinguished itself. This was the trick drill platoon of the Women's Reserve Battalion, Marine Corps Base, Quantico which gave special performances for visiting dignitaries and appeared at many social and military functions. Organized in April 1944, the trick drill team was complimented on a letter from Brigadier General Archie R. Howard. In his commendation message, he noted particularly that the excellence of the team's work "has been attained by personal sacrifice of many liberty hours, in as much as all time devoted to instruction has been entirely voluntary and in addition to regular assignments."(179)

Personalities in the WRs

Many Women Marines created news from the time of their enlistment. For example, there was Marine Private Natica R. Macy who was living in Bermuda when she heard about formation of the Marine Corps Women's Reserve. She immediately hopped an ancient Greek freighter bound for the States, wound up being the sole woman aboard the vessel and did a man-size load of chores for her passage--including painting woodwork, calking decks, mending socks, and standing watch. "It was the first available transportation," she explained simply.(180)

The first full-blooded Indian recruit was Private Minnie Spotted-Wolf of Heart Butte, Montana, who enlisted in July 1943. Despite her slight 114-pound frame, the 20-year old Indian girl had driven a two-ton truck, cut fence posts, erected bridges, and broken horses on her father's ranch. Her comment on boot camp training: "Hard, but not too hard."(181)

Two other Marine representatives of the Indian race were Privates Celia Mix, of Benton Harbor, Michigan and Viola Eastman, from Pipestone, Minnesota. Celia's brother, Corporal

John Mix, was a Marine who had made headlines by winning the Air Medal for heroism in saving the life of a fighter pilot who had crashed into the South Pacific Ocean. Celia was on her way to setting a record of her own: she went all the way through boot camp without receiving one restriction or demerit.(182)

The first Woman Marine from Puerto Rico was Private Norma Frances Aran, who had formerly been a civilian employee of the Army. She reported her enlistment came as no surprise to her employer, commenting: "the Army had known for a long time that my heart was with the Navy." Another who deserted the Army to join the Marines was Virginia Carter, whose father, Master Sergeant William S. Carter, stationed at Fort George G. Meade, Maryland, was a 26-year veteran of Army service. Virginia saw 2,000 persons march in a military parade held on the base in her honor, including WACs, an Italian unit, and three bands.(183)

Another Woman Marine who created news was Captain Lily S. Hutcheon, one of the original 19 ex-WAVES, who in May 1944 became the first woman Judge Advocate in the history of the Marine Corps, assigned to duty at Camp Lejeune, North Carolina. Before receiving her commission, she had been employed in the legal department of a San Francisco oil company.(184)

There were a number of sisters who enlisted together in the Women's Reserve and several sets of identical twins. The latter included Betty and Bonnie Jernigan, from Sparta, Tennessee, first set of twins to enlist in the MCWR; Madelene A. and Irene A. Spencer, who were on recruiting duty in Philadelphia, Pennsylvania; and Mary E. and Martha H. Taylor, who were stationed in the Dallas Procurement District.(185)

Although a number of American parents routinely faced the departure of two members of their family when sisters left for recruit training, a novel situation occurred to one Philadelphia family early in 1944. They "lost" two members to the Marines when Private Ruth H. Whiteman enlisted and took her Doberman Pinscher, Eram Von Lutenheimer, along to join the Marine Corps War Dog Detachment at Camp Lejeune.(186)

And of course many women were in a sense Marine Corps women long before their enlistment, by virtue of having fathers, brothers, or uncles in the service. Among these was Mary Cleland Fordney, a granddaughter of Major General Ben Hebard Fuller, 14th Commandant of the Marine Corps, and whose father, Colonel Chester L. Fordney, was Officer-in-Charge of the Central Procurement Division of the Marine Corps. Another was Eugenia Dickson Lejeune, daughter of the late Lieutenant General John A. Lejeune, Commandant from 1920 to 1929. She underwent basic training at Camp Lejeune, named in honor of her father. She and Second Lieutenant Fordney were both members of the seventh officers' class, graduated

15 November 1943. Also a member of this same class was Second Lieutenant Patty Berg, golf professional and queen of the fairways, who had enlisted in the Marine Corps at a women's services recruitment/golf tourney and whose birthday, 13 February, was the same as the anniversary date of the Women's Reserve. She purposely left her clubs at home for the duration, declaring she wanted to be "strictly a Marine," and was assigned to recruiting duties in the Eastern Procurement Division.(187)

Another woman with Marine Corps interests at heart and at home was Private First Class Aline Bernice Shelton, of Phoenix, Arizona, the ninth member of her family to be a Marine. She enlisted as soon as it was possible for her to do so: on her 20th birthday. One of her brothers was a prisoner of war and several others had hazardous Pacific duty. Helen Kemp Devereux, 20-year-old niece of Lieutenant Colonel James P. Devereux, still a prisoner of the Japanese in 1945, also enlisted in the MCWR. Petite 95-pound Helen also had three brothers, six other uncles, and one cousin, all of whom were Marines. There was even a tradition of women Marines in Helen's family: one of Colonel Devereux's sisters had been a Marinette in World War I!(188)

A Woman Marine who followed her mother's footsteps by joining the Marine Corps was Private First Class Grace E. Mather, believed to be the only WR able to claim the honor of being a second-generation Woman Marine. Her mother was the former Corporal Helen M. Dunn, who had worked in the muster roll section at Washington during World War I.(189)

Other WR Interests and Activities

In addition to regular duty assignments, varied wartime activities of the Women Reservists ran the gamut from participating in christening ceremonies of the "Lady Leatherneck," a Marine Corps transport plane--to blood donations, visiting the wounded at Army and Navy hospitals, volunteering in extra civic duties, public appearances, and war bond rallies. As for buying bonds--that was routine. Ninety-four percent of the women bought them every month.(190)

A number of women with a studious bent also enrolled in correspondence courses given by the Marine Corps Institute, with Spanish being the most popular choice. Other courses taken by the women included stenographic lessons, short-wave radio theory and mathematics, aviation, motion picture sound technicians's course, a current event analysis, "The Pacific World," and--with an eye toward their own destinies--cooking and sewing courses. More than 10 percent of the women took advantage of Marine Corps Institute instruction.(191)

Decorations Awarded Women Marines

The highest award made to a Woman Marine as a result of World War II service was the Legion of Merit, awarded the two wartime Directors, Colonel Ruth Cheney Streeter and Colonel Katherine A. Towle. Two women, Majors Helen N. Crean and Marion Wing, received the Bronze Star Medal. Letters of Commendation and Commendation Medals were presented to more than 30 enlisted women and officers of the Women's Reserve. As a result of World War II service, Women Marines were eligible to wear the Good Conduct Medal, American Campaign Medal, World War II Victory Medal, and Asiatic-Pacific Campaign Medal.(192)

VIII. Hawaii Duty

Scuttlebutt was rampant during the summer and early fall of 1944 concerning the possibility of Women Marines being sent overseas for the first time. For more than a year Congress had debated the issue, which involved amending the Naval Reserve Act of 1938 to allow women members of the naval service to serve outside the continental limits of the United States. Although originally there had been serious objection voiced in the House, (193) suddenly it looked as if the Senate would take favorable action, spurred on by the Senate Naval Affairs Committee, on its companion bill. As yet there had been no official word as to whether Women Marines would be needed in overseas billets, such as Pearl Harbor, to release additional men for combat duty in the final push to victory. A number officers back from the Pacific, however, including Lieutenant General Holland M. Smith, had "mentioned the matter informally." (194) An estimated 5,000 naval servicewomen were needed immediately in Hawaii alone, Secretary of the Navy James V. Forrestal had told Congress.(195)

On 13 September, the Senate passed the bill which was subsequently adopted by the House and signed into law by the President on 27 September 1944. This bill (Public Law 441, 78th Congress) modified the existing Department of Navy regulations and thus permitted "female Naval personnel to serve on a volunteer basis anywhere within the western Hemisphere, including Alaska and Hawaii."(196)

Excitement ran high among members of the Women's Reserve when it was apparent that some of them would be going overseas. As for the Women's Reserve Director, she knew that passage of the legislation would precipitate a host of new problems and procedures and, in fact, had already anticipated some of them. Back in July she had submitted tentative policies for selection of personnel in the event overseas service of women was authorized and needed by the Marine Corps.(197)

The Marine Corps contemplated sending one group of women to Pearl Harbor to form the Women's Reserve Battalion attached to the Marine Garrison Forces, and another smaller detachment to the Marine Air Station at Ewa. To find out exactly what the conditions were in Hawaii for this new duty, Colonel Streeter and Major Marion B. Dryden, the senior Women's Reserve personnel officer in the Division of Aviation, set out on a fact-finding mission on 13 October. Flying to Hawaii, they inspected the area and talked with both Marine and Navy officers to learn the principal job categories of men to be released for combat, what quarters were available for the women or could be converted for their use, the existing chain of command, recreation opportunities, regulations and procedures followed by service personnel on the island, and related matters. Upon their return they recommended assignment of two detachments of Women Marines, as originally planned. Although the two new units were to be administered separately by their own women commanding officers, ultimately both were to come under the final jurisdiction of the 14th Naval District.(198)

Selection of Women for Overseas Duty

Every woman who volunteered for Hawaiian duty was carefully screened. Selection was based partly on length of service in the Marine Corps, but primarily on the job classification of the men to be relieved. The principal job categories included clerical, communications, quartermaster, telephone operators, mechanical, motor transport, and radio operators. In the group chosen were a number of women whose brothers had been killed in action, had become prisoners of war, or who still had relatives on the fighting front.(199)

A sense of responsibility, maturity, adaptability, and emotional stability were the chief personal characteristics considered necessary in the candidates.

In addition, other necessary qualifications were at least six months' service on active duty, excluding training time; good health, conduct, and work records; and freedom from any form of dependency which would require their presence at home. This was particularly important since overseas duty was for a two-year period, and leave or furlough to return to the States was to be authorized only under extreme emergency.(200)

Because of the wording in the enabling legislation, it was necessary that a woman naval reservist specifically volunteer for the overseas duty; she could not be ordered to it. In processing applications, careful consideration was given to a woman's motivations for volunteering. Any notions of excitement or so-called glamour attached to the duty were sharply dispelled by the Marine Corps. A memorandum to Colonel Streeter from Brigadier General Waller, now Commanding General, Marine Garrison Forces in Hawaii, focused attention on the facts: "All WRs should be informed this duty here is not glamourous--just hard work," he advised. "They will be under more restrictions than at home and their working and living conditions may not be as good....They will be closer to the war, they will see ships damaged in combat and they will see and meet many men who have recently been in combat...."(201)

One recommendation which had come from the Canadian Women's Army Corps and which became implicit in Marine Corps' thinking was that a volunteer should have demonstrated she can "maintain good adjustment when her work is dull or monotomus or when she has to work under pressure or conditions of strain."(202) This, too, was an important consideration. All operations in Hawaii worked a seven-day week, and although regular schedules of half-day or one full day a week liberty were arranged, in times of crisis an installation not infrequently worked around the clock until a particular priority job was completed.

Advance Party

On 2 December, an advance party of four officers flew to Hawaii to make preliminary arrangements for the organization and reception of the battalion of approximately 100 housekeeping personnel which would follow shortly, as well as the eventual full complement of nearly 1,000 Women Marines. Members of this initial group were Major Marion Wing, who was one of the original 19 ex-WAVES and who was assigned as Commanding Officer at Pearl Harbor; First Lieutenant Dorothy C McGinnis, Adjutant at Pearl Harbor; First Lieutenant Ruby V. Bishop, Battalion Quartermaster; and Second Lieutenant Pearl M. Martin, Recreation Officer.(203)

They were followed by the advance party for the aviation unit, consisting of Captain Helen N. Crean, Commanding Officer at Ewa; First Lieutenant Caroline J. Ransom, Post Exchange Officer; Second Lieutenant Constance M. Berkolz, Pearl Harbor Mess Officer; and Second Lieutenant Bertha K. Ballard, Mess Officer at Ewa.(204)

Staging Area and Arrival in Hawaii

In January 1945, the first volunteers for overseas duty were transferred to a staging area which had been established at the Marine Corps Base, San Diego, California. Here they were given a short but intensive physical conditioning course including qualification swimming, drill, and calisthenics. The women also learned to ascend and descend, with a full 10-pound pack on their packs, the cargo net of a ship mock-up and how to jump properly into the water from shipboard, in event they should be aboard a transport that met with enemy attack en route to its destination.(205)

They were also given physical examinations, inoculations, a brief review in certain phases of Marine Corps administration and organization, inspection of uniforms and gear, lectures about the people of Hawaii, Allied insignia, safeguarding of military information, procedures on shipboard, and a final screening.(206)

On 25 January, the first contingent of 160 enlisted and five WR officers sailed from San Francisco aboard the S. S. Matsonia.(207) Dressed in winter greens and trench coats and carrying blanket rolls, the women marched up the gangplank and aboard ship in column and proceeded to assigned quarters. Two days out at sea they changed to summer service uniforms, which would be the uniform of the day for the length of their Hawaii duty. En route, the women staged their own entertainment show and performed the long-familiar ritual of policing their own area.(208)

The Matsonia arrived in Honolulu on the morning of 28 January. Captain Marna V. Brady, Officer-in-Charge of the voyage, who was assigned as Battalion Executive Officer, led the first enlisted women to disembark, was greeted with the customary Hawaiian lei and kiss.(209)

To one Woman Marine the overseas assignment was a trip home! She was Corporal Alice M. Philpotts, who came to the States from her home in Honolulu, only a few miles southeast of Pearl Harbor, to enlist in the Marines and who had been at Pearl Harbor on the date of the attack.(210)

For the rest of the women, however, Hawaii was as much a novelty as they were to the islands. From the crowds assembled at dockside there were cheers by residents, civilian war workers, and servicemen; tons of colorful floral leis; the flash bulbs of photographers; and amazed expressions of male Marines who had never seen a Woman Marine before. The Pearl Harbor Marine Barracks band alternated between the soft tune of the traditional Aloha Oe, and the martial strains of the Marine Hymn and official March of the Women Marines. The women were quickly dispatched to their new homes: the majority to the Moanalua Ridge area, where the Women's Reserve Battalion occupied a former Seabee area adjacent to the Marine Corps Sixth Base Depot and Camp Catlin; the air group to the nearby Marine Air Station, at Ewa, on the island of Oahu.(211)

On the streets people stopped to smile and wave at the women. When they filed into the mess hall for their first meal, a little dog growled at the unusual sight of green-uniformed women but quickly subsided when a messman exclaimed: "Knock it off, Taffy. After all, they're Marines, too!"(212)

In Hawaii, as in the States, the Women Marines replaced men not only in office jobs but in specialized work as well. They stood night watches in communications and other duty assignments, and generally worked the same hours as the men.

At Pearl Harbor, the women ran the entire motor transport section, which served approximately 15,900 persons a month including liberty busses, work detail trucks, and jeeps--all with a perfect safety record. A total of 33 vehicles were operated by the section, working a 24-hour-a-day schedule.(213)

Women Marines assigned to the air station found that duty was much like aviation jobs anywhere. They had little trouble orienting themselves, as general surroundings and living conditions were not much different from those of mainland air stations. In fact, more than one-third of all the women had previously been stationed at Cherry Point.(214)

New detachments from the States of approximately 200 enlisted women and 10 officers arrived every other week. By the time the fourth group had arrived, the WR Battalion was a smoothly-running outfit. The appearance of the quarters area had measurably changed for the good. Shrubs, small trees, and blooming plants appeared to have sprung up almost overnight. Detailed by their Brigade Headquarters, the Seabees had done the construction and renovation work on the administration building, barracks, mess-hall, and laundry. The women pitched in on landscaping the area and helping the Seabees. As one Marine good-naturedly grumbled at the time: "The Army has its WACS; the Navy, its WAVES; the Coast Guard, its SPARS; and the Seabees, OUR Marines!"(215)

From the day of arrival, the women were in such social demand that requests to entertain them had to be screened through an enlisted women's council which decided, on behalf of all the WRs, just what unit of the infantry, artillery, cavalry, Seabees, engineers, tanks, etc., could be granted acceptances. On several occasions, weekend air hops were arranged so that the women could visit the Marine detachments on the islands of Hawaii and Maui.(216)

And of course, the inevitable happened. In May 1945, a Marine combat correspondent wrote a little history of his own when he married a Woman Marine. The principals in this first all-Marine overseas wedding were Sergeant Dorothy Jeanne Crane, a photographer on duty with the Marine Garrison Forces, and Staff Sergeant Robert T. Davis. Although the bride wore traditional white bridal gown and finger-tip veil, the tropical touch was apparent in the white orchid tiara and pikaki leis in her bouquet. Most of the guests were men and women Marines. One of the few civilians present was Representative Margaret Chase Smith, only woman member of the House Naval Affairs Committee, who had helped draft legislation for the overseas bill and who was then in Hawaii on an investigative tour.(217)

Approximately 1,000 women saw duty with the Marine Garrison Forces at Pearl Harbor and at the Marine Corps Air Station at

Ewa, and nearly all agreed that overseas duty made them feel they were taking a more active part in winning of the war. Announcement of V-J Day in September 1945 brought their expected two-year duty stint to an abrupt halt. The first group of women left Hawaii early in December 1945, in time to make words of the then-popular lyric "I'll be Home for Christmas" really ring true. The rest returned Stateside the following month. And despite the sudden cessation of hostilities and the victorious war news, many Women Marines still couldn't help being disappointed at having such "wonderful duty" cut short.(218)

IX. Demobilization

With the ending of hostilities on 14 August 1945 and V-J Day on 2 September, all recruiting for the Women's Reserve was halted although women in training classes continued until completion and were assigned to duty. Actually, since the Marine Corps had reached its original goal of approximately 18,000 enlisted and 1,000 officers in the summer of 1944 recruiting had been deliberately slowed down and the comparatively small mumber of women who had entered the MCWR toward the end of the war had been largely replacements for normal attrition.(219)

Demobilization plans moved ahead rapidly and efficiently, under an "Adjusted Service Rating System" of points, similar to that for the men. Computation of these discharge credits was the same for the women as the men. However, since the Women Marines were able to earn credits only by length of service (not having been in combat or having dependent children) the number of credits required for discharge was originally set at 25, compared to 85 for the men.(220)

The terminal date originally set for the Women's Reserve was 1 September 1946. All women were to be discharged by that time. Or, as the Headquarters Marine Corps bulletin stated: "Officers and enlisted personnel of the Marine Corps Women's Reserve are now being rapidly separated from the service because of the fact that the entire Women's Reserve is to be demobilized by 1 September 1946."(221)

With the Marine Corps candor the women had learned to love and respect, the Commandant added:

"It was with some hesitation the the Marine Corps admitted women to its ranks in February 1943, but during the intervening years they have made a most valuable contribution to the Corps...

"As the time comes to release them, I am reminded again of the important part they have played in support of our combat Marines while the actual fighting was in progress...

"I wish to express to the members of the Women's Reserve the appreciation of the Marine Corps for the valuable contribution they have made for its success. They have performed their duties in a manner that evokes the admiration and praise of their fellow Marines; and their conduct and appearance, both on and off duty, have been exemplary and a source of pride to us all."(222)

A two-week "Rehabilitation School" for Women Reserve officers and noncommissioned officers was set up at Headquarters and at Camp Lejeune. It provided information about rights and benefits under the Veterans Administration and G. I. Bill of Rights so they could properly advise the women under their direction who were being discharged to civilian life. In many cases letters of recommendation were written to former and prospective employers. For women desiring high school credit, letters were written to hometown high schools requesting credit on the basis of recruit training and experience in the Women's Reserve. Colleges were also contacted for brochures and information regarding entrance requirements. A poll taken by one group of women being discharged showed that the most popular plans for the future, in order, were: new employment, education, old employment, housewife, plans "indefinite," and civil service.(223)

Computation of Credits

All credits for demobilization of the women were based on their length of service and computed from 1 September 1945, the control date. The credits were progressively reduced until, on 1 July 1946, they became zero. A woman having 25 credits in September 1945 was eligible for immediate discharge; by the following January, those with 18 credits could be discharged, and so forth. The schedule of credit reductions and their effective dates, spelled out by Letter of Instruction 1110, were:(224)

Effective date	Credit
1 September 1945	25
1 November 1945	20
1 January 1946	18
1 February 1946	17
1 March 1946	16
1 April 1946	13
1 May 1946	8
1 June 1946	4
1 July 1946	0

Regardless of actual number of credits, immediate discharge upon request was authorized for all women 38 years of age or more (later changed to 35 years); or for a married woman if her servicemen husband had been discharged. Married women with a year's active duty were also granted immediate release upon request if their husbands were in the country, whether discharged or not, and regardless of branch of service.(225)

Newspaper accounts in November 1945 quoted the Commandant as saying that the Marine Corps Women's Reserve would be reduced from 18,000 enlisted and 1,000 officer strength to 2,638 and 200, respectively, by 30 June 1946 and that the organization "will completely vanish from the picture by September of next year."(226)

The MCWR was reduced to two-thirds of its peak strength by 7 December 1945. On that date its original Director, Colonel Streeter (twice-promoted--to lieutenant colonel on 22 November 1943 and to colonel on 1 February 1944), resigned to be home for her three sons, all returning from overseas duty. She was succeeded by Lieutenant Colonel Katherine A. Towle, of Berkeley, California, who assumed the Directorship and its rank of colonel on the fourth anniversary of Pearl Harbor.(227)

Another long-familiar institution that went about this time was the Women's Reserve Band. This was officially disbanded at Camp Lejeune on 28 November 1945 when 26 members were discharged under the credit provisions of Letter of Instruction 1110. The remaining 21 members were reassigned in the Women's Reserve Battalion. Low-point band members met at the Separation Company on the morning the majority of band members were being discharged and serenaded them. A gold identification bracelet bearing the woman's full name and with the inscription "Thanks from Camp Lejeune" was presented to each member of the band.(228)

Strength at End of the War

In August 1945, two and a half years after formation of the Marine Corps Women's Reserve, there were approximately 17,640 women and 820 officers on duty, or a total strength of 18,460. There were 28 units headed by women commanding officers, plus 17 smaller units, and additional women assigned to specialist duties, such as recruiting.(229)

Line units included Women's Reserve Battalions at Henderson Hall; Quantico; Camp Lejeune; Parris Island; San Diego; Camp Pendleton; and Pearl Harbor; the School Detachment at Camp Lejeune; and Women Marine companies at San Diego; Department of Pacific, San Francisco, the Navy Yard at Mare Island, California; and in Washington, D. C.(230)

Aviation units included those at Cherry Point; Quantico; Parris Island; El Toro; Miramar; El Centro; Santa Barbara; Mojave; Ewa, Hawaii; and Eagle Mountain Lake, Texas.(231)

The four quartermaster units to which Women Marines had been assigned were: Depot of Supplies in Philadelphia; South Annex, Norfolk; Camp Elliott, California; and Depot of Supplies, San Francisco.(232)

Women were also stationed at the four procurement districts: Eastern, at Philadelphia; Southern, in Atlanta; Central, in Chicago; and Western, at San Francisco.

Monthly Quota for Demobilization

The demobilization procedures for male personnel were paralleled for the women: officers were released to inactive duty and enlisted women were discharged. Beginning in October 1945, control of the release of Women's Reserve personnel was in January 1946, the rate of discharge of enlisted women was set at a minimum of approximately 1,100 each month. This figure was selected because it would enable a gradual but steady reduction so that final disbandment on 1 September 1946 could be accomplished smoothly, without any operational inefficiency.(234)

When the number of Women Marines at any post or station became fewer than 100, that particular Women's Reserve activity was disbanded or transferred administratively to an adjoining unit. All small and scattered units were disbanded as quickly as possible, with the exception of a few "high priority" units such as district rehabilitative and recruiting centers. Final disbandment dates were also set for each post and station. When it came time for the post to be officially closed, personnel still on duty were discharged, if eligible, or transferred to another station.(235)

Morale was high despite the inevitable "anticlimactic" feeling that comes with mopping-up operations. A big determinant was the fact that discharge of those eligible moved along at a satisfactory rate. And there were plenty of activities to keep a Woman Marine busy until her discharge papers came through. At one base a small but enthusiastic group in a beginning French class completed the rest of its lessons after its instructor was discharged.(236) In December, Women Reserves at the Marine Corps Air Depot at Miramar entered the 11th Naval District contest for suggestions on the conversion of military clothing to civilian.(237) At Henderson Hall a three-part lecture series, "You Are Prettier Than You Think" was given by Mrs. Hanna Sherman, formerly associated with a leading New York cosmetic firm. Lectures dealt with readjustment to civilian life--including makeup and restyled hair, suggested plans for a new wardrobe, and job-hunting tips to ex-Marines turned business girls. At the same base, a large number of women officers volunteered for a day's mess duty during the holidays as a Christmas present to the enlisted women.(238) And at the big training base, Camp Lejeune, a collection of songs sung by Women Reservists during their service in the Marine Corps was compiled and distributed to each girl on the day of her discharge.(239) A statement made in January 1946 by one Woman Marine officer was typical: "Standards of work and attitude among the Women Reservists have continued to be high. There is still the desire to serve faithfully and well."(240)

Separation Centers

When the demobilization process began, commanding officers were authorized to discharge the women at their duty stations. To further expedite matters, however, four separation centers were set up in October to process discharges of the women. These were: Henderson Hall and Camp Lejeune on the East coast; and the Marine Corps Base at San Diego and Marine Corps Air Station, El Toro on the West coast. A unique arrangement existed at San Diego. By special permission from Headquarters, the Women's Reserves separation activities there became part of the male First Separation Company. Women's Reserve rehabilitation personnel were available to the women and a Woman Reserve officer served as Adjutant and Officer-in-Charge of the women attached to the company. The arrangement worked well, and the women themselves like being discharged with the men, as it gave them the feeling they were very much a part of the Marine Corps even until the end.(241)

Besides the four major separation centers, small units and the detachments at San Francisco and Parris Island were authorized to discharge their own personnel. Women returning from Hawaii were transferred to the separation center nearest their homes: those from the East or Middle Eastern section of the country to Henderson Hall or Camp Lejeune; and persons from the West or Southwest to either San Diego or El Toro.(242)

Last Days of the Wartime Reserve

By 1 June 1946, the Women's Reserve activities had been disbanded at these major stations: Camp Lejeune (Women's Reserve Battalion); Parris Island; and the Depot of Supplies, Philadelphia. By 1 July, WR units had been de-activated at Quantico; Camp Lejeune (Women's Reserve Separation Center); Camp Pendleton; and Marine Corps Air Depot, Miramar.(243)

The only Women's Reserve units remaining until the 1 September terminal date were those at Henderson Hall and Cherry Point on the East Coast; and El Toro and Department of the Pacific, San Francisco on the West. Women Reservists remaining on duty between 1 July and 1 September were, with a few exceptions, volunteers. The strength of the organization had been reduced to approximately 1,000 women by 1 July, and this number was gradually decreased in the final two-month period. During these last months, the majority of Women

Marines still in uniform were stationed at Henderson Hall, on duty at Marine Corps Headquarters. (244)

Upon termination of the Office of the Director of the Women's Reserve, on 14 June 1946, and prior to her return to civilian life, Colonel Towle, the second Director observed:

"General morale during demobilization has been gratifyingly high. Part of this has been due to the definite stand the Marine Corps itself has taken from the beginning on the Marine Corps Women's Reserve demobilization, particularly in setting and maintaining 1 September 1946 as the terminal date of the wartime Women's Reserve. It has been a goal to work toward, and Marine Corps women have never had the uncertainty and confusion concerning demobilization which have occurred in some of the other women's services because of the shifting of dates and changes in policy...."(245)

X. Overview

Since time immemorial military discipline and regimentation have been routine and accepted for men. This type of training for women, on a large scale, was a totally new concept in World War II. This immediately raises several questions.

First: What civilian background or personality syndrome seemed to be the best determining factor in developing military aptitude in a woman?

Second: Since women are generally considered the arch individualists of the race, how did they react to military discipline and a situation in which much of their life was planned and provided for, with less opportunity for individuality?

Third: Did military experience in the MCWR have any lasting benefit on the individual woman, besides giving her a feeling of actual participation in winning the war?

Civilian Background -- Actually, no one particular civilian occupation seemed to prepare a woman for the unique responsibilities of military life. Similarly, no special personality trait seemed to guarantee success. In general, it can be said that qualities which helped a woman become a good Marine included: a well-balanced personality; skill and efficiency in her work; adaptability to new ways; energy; emotional stability; sense of responsibility; and promptness in getting things done and carrying out orders.(246)

Although the quality of leadership is difficult to define, it was found that the best officers were those with strong personalities, who were able to command respect and to motivate others effectively. Other important qualities were responsibility, fairness, good judgment, understanding and respect for others, professional competence, good personal appearance, self-confidence, and emotional stability.(247)

Reaction to the Military Life--As was true with all the women's services, there was naturally a small percentage of misfits who would be unhappy in any group situation, civilian or military. Most of the women, however, adjusted quickly and well to military life and its inherent discipline. Surprised drill instructors often found that the women seemed to snap into the esprit and precision of close-order drill, for example, even faster than the men. Furthermore, in a survey taken in the first months of the Marine Corps Women's Reserve, it was found that a number of members complained there was "not enough drill, not enough regimentation, too much like civilian life. Would enjoy more militarism...."(248) Such changes in the direction of a more military atmosphere were effected in July 1943, when all basic training for the women was centralized at Camp Lejeune, where it remained throughout the rest of the war.

Personal Benefit--Probably no woman has ever worn the Marine uniform without gaining something new of permanent value to herself from the traditional Marine Corps insistence on order, organization of work, strict responsibility down to the last detail, getting things done right in a minimum of time, self-discipline, pride, self-confidence, and flexibility. By sheer necessity in the MCWR, most women learned new habits of first things first, overcome indecision, take immediate action, and tackle unfamiliar (if not downright distasteful) chores and get them out of the way. Then, too, there was valuable personal training in doing those things which a woman at the onset might have felt she didn't have real confidence she could do--such as supervising the work of others.(249)

Like all good Marines, the women learned to improvise, to adjust quickly and make the best of the situation--whether it was emergency transportation to a new duty station in miserable cattle trucks or entertaining the Director graciously in their own barracks because of lack of conventional recreation facilities.(250) As Colonel Streeter herself once noted in a discussion about the effect of women being in the service: "Our members do not lose any of their womanly qualities, but their health is improved and their point of view changed by the military discipline they undergo. It is not an easy life, and I am constantly surprised at their loyalty and endurance...."(251)

Educational Background

Women Marines represented a variety of background and training. An analysis of their educational background showed that out of 21,050 women surveyed, a total of 13,824, or approximately 88 percent, had completed high school. Since one of the requirements for enlistment was two years of high school, this percentage was not surprising. An additional 4,619 had attended college.(252) (See Appendix B)

Regional Pattern of Enlistments

A geographical study of the residence of Women Marines revealed that 80 percent came from 18 states, while the other 30 states supplied only 20 percent. The states of California, New York, Pennsylvania, Massachusetts, Illinois, Michigan, and Ohio supplied 12,769 Women Marines, or more than half the total number. The fact that six of these states which supplied the most women were also among the seven most-heavily populated states during the war years shows a definite relationship in most cases between a state's population figure and the number of women who joined. There are, however, several interesting notes on the representation from the various states. For example: although the state of Texas ranked sixth in population, it was fourteenth in the number of women who enlisted in the Marine Corps. California during the war years ranked fifth in population, but first in the number of Women Marines enrolled. And Massachusetts, which ranked fourth in the number of women enlisted, was eighth in total population.(253) (See Appendix C)

Regionally, the highest proportion of Women Marines came from the Middle Atlantic states, the Central states, the Far West, and the Great Lakes states, in that order.(254) From the beginning, it was more difficult to interest women from the Southern states in enlisting. Recruiting officials in this section had to work particularly hard to fill their monthly enlistment quotas. As Director Streeter once wrote to an officer in New Orleans: "We at Headquarters realize the difficulties faced by Procurement offices in the deep South and appreciate the efforts which you are making to overcome sales resistance in that area. Service in women's military organizations is a newer idea in the deep South than in other parts of the country and more at variance with their customs and traditions. However, I hope they may in time come to see its importance...."(255)

In addition to maintaining a steady flow of enlistments to meet the needs of both the enlisted and officer training programs, it was also necessary to meet recruiting quotas designed to provide the Marine Corps with women representing a balanced cross-section of the country.(256)

Composition of Reserve by Age and Test Scores

Primarily, the Women's Reserve was composed of women in their early 20s. (See Appendix D) At the end of the war more than 60 percent of the officers were in the 20-24 age group. When recruiting opened in February 1943, an age range between 20 and 50 was allowed for officer candidates. This, at least theoretically, made it possible to achieve a fairly balanced ratio between young officers and the more professionally experienced, older ones. However, later that fall when the officer classes were composed primarily of enlisted women and, in general, closed to persons from civilian life (with the exception of a small number of highly-qualified candidates), the age distribution shifted toward the younger side and remained there. This did not cause any great problems, although a better balance would have been considered desirable. There was also the feeling that, particularly in the case of officers assigned to troop duty, sometimes relatively young lieutenants carried a heavy load of responsibility for their age that far outweighed commensurate duties in civilian life. (257)

The median age of all Women Marines was 24.05 years.(258)

In any organization such as the Women's Reserve, it is desirable to have a range of abilities represented by its members, since there are a number of fairly routine jobs which have to be done.(259) Even in a wartime situation, if there are too many over-qualified people in comparison to the work that has to be done, it is sometimes difficult to escape the problem of "under-assigning" personnel and this may result in dissatisfaction and lowered morale. On the other hand, the enrollment of high-quality personnel in an organization usually contributes considerably to its ease of administration and general lack of serious disciplinary problems. It was found that the frequency of minor disciplinary problems and maladjustments which occurred to women in the lower General Classification Test groups (i.e., under 75 and in the 75-88 GCT range) was "out of all proportion to their actual numbers."(260) In the Women's Reserve, the vast majority of personnel ranked 89 or higher in the standard aptitude test which measures basic intelligence and innate learning ability. (See Appendix E)

Recruiting Results and Media Used

Recruiting figures showed that a total of 23,145 women were enlisted during the war, and that only 965 ever held commissions. The Women's Reserve achieved its recruiting goal of approximately 18,000 enlisted and 1,000 officers several weeks earlier than its target date of July 1944. Thus, during the months of July and August 1944 there were virtually no Women Marines accepted, to avoid the potential problem of supply in excess of number of billets available. In September 1944, recruiting was reopened to provide replacements for Women Reservists volunteering for duty overseas.(261)

The most successful media used to aid recruiting were, in order: radio, newspaper publicity, posters and outdoor advertising, and movies.(262) From the onset it was recognized that a good public information program was essential to win continuing public support and to attract the high caliber of personnel desired for the Women's Reserve. The Division of Public Information of Marine Corps Headquarters released information to all media about practically every aspect of the program. All official MCWR publicity was kept on a high plane carefully avoiding the come-on of glamour, foreign assignments, or the like. The women were told their jobs were not glamourous but hard work, just as the individual Marine's job in the Pacific was not glamourous.(263)

Not unexpectedly the most successful type of appeal was patriotism. After V-E Day, 7 May 1945, it became harder to get good-quality Women Reserve candidates, since the patriotic influence was no longer as strong. A survey was conducted by the psychiatrist at the Camp Lejeune Recruit Depot when it was operating at full strength in the winter of 1944. He did not select a particular group, but simply asked questions of the first 1,000 recruits who were taking their physical examinations at that time. They were told: "We know that the desire to serve your country was your primary reason for enlisting in the MCWR. What would you consider your secondary reason?" The answers were as follows:(264)

	<u>Positive reasons</u>	
1.	Because they had men in the service.	350
2.	Because there were no men in their families who could serve.	60
3.	For revenge: their men had been killed.	40
4.	For adventure.	150
5.	To benefit themselves.	150
		750
	<u>Negative reasons</u>	
1.	To "get away from something" such as a distasteful job, family difficulties, etc.	250
	TOTAL	1,000

Other surveys conducted indicated that the majority of women were proud to belong to the Marine Corps. They joined because they wanted to do a job and get the war over. They picked the Marine Corps because to them it stood for the highest in tradition, ability, and accomplishment.(265)

Overall Distribution by Rank

The original request to the Secretary of the Navy for authority to organize a Marine Corps Women's Reserve asked for a distribution in the ranks of officers to conform to that for enlisted Women Reservists the same as that authorized for enlisted men of the Marine Corps. This was approved as follows: (266)

Officers	
Major	1
Captain	35
1st Lt	35% of total authorized commissioned
2nd Lt	Balance

Enlisted	
1st Pay Grade	3.68[illegible]%
2nd Pay Grade	5.73[illegible]
3rd Pay Grade	7.714
4th Pay Grade	12.605
5th Pay Grade	23.95[illegible]
6th Pay Grade	37.221
7th Pay Grade	9.087
	100.000%

This initial distribution in rank of officers was modified on 25 November and the following distribution was authorized:

Colonel	1
Lieutenant Colonel	1
Major	30
Captain	100
1st Lieutenant	240
2nd Lieutenant	Balance

This also proved inadequate and was later modified. Final distribution in rank was established in January 1945 following approval by the Secretary of the Navy of a request by the Commandant for a plan that better met the needs of the service. This new distribution provided for:

Colonel	1
Lieutenant Colonel	4
Major	36
Captain	200
1st Lieutenant	400
2nd Lieutenant	Balance

"VIP" Statements about Wartime Reserve

How effectively Women Reservists lived up to their wartime recruiting slogan "Free a Man to Fight" was expressed by President Franklin D. Roosevelt in the following message sent them on the first anniversary of the Women's Reserve, 13 February 1944:

"The nation is as proud of you as of your fellow Marines-- for Marine women are upholding the brilliant traditions of the Corps with a spirit of loyalty and diligence worthy of the highest admiration of all Americans. You have quickly and efficiently taken over scores of different kinds of duties that not long ago were considered strictly masculine assignments; and in doing so, you have freed a large number of well-trained, battle-ready men of the Corps for action...."(267)

Declared General Thomas Holcomb, the early wartime Commandant: "There's hardly any work at our Marine stations that women can't do as well as men. They do some work far better than men....What is more, they're real Marines. They don't have a nickname, and they don't need one. They get their basic training in a Marine atmosphere, at a Marine Post. They inherit the traditions of the Marines. They are Marines."(268)

Because of the recruiting slogan adopted by the Marine Corps when it began enlisting women, the question of how many men were so freed was naturally a subject of interest. The peak strength of the Women Reservists, slightly less than 19,000, approximates a Marine Corps division. Therefore, a remark always treasured by the women was the statement made by General Alexander A. Vandegrift, the second wartime Commandant, who said they could feel responsible for putting the 6th Marine Division in the field; for without the women filling jobs throughout the Marine Corps there would not have been sufficient men available to form that division.

On another occasion he observed: "With quiet assurance, and without fanfare, they have learned quickly all tasks assigned to them. In doing so, they have proved themselves so versatile and so adept that thousands of men were released earlier than had been hoped, to take part in the great Pacific drive which will continue on its relentless way....I have been equally impressed with the manner in which they have taken the traditions of the Corps to heart. They have developed an esprit worthy of the admiration of the most thorough-going veteran in our ranks...."(269)

On the Women's Reserve second anniversary, 13 February 1945, General Vandegrift declared:

"Just two years ago, the United States Marine Corps called on the women of this country to help it meet the severest test in Marine Corps history....

"You responded generously. Thousands of you came forward of your own free will to join the Corps.

"Today you number from approximately one-third to one-half of the post troops at representative Marine posts and stations; and, as might be expected from the type of work to be performed, your services have been particularly in demand at Headquarters, where you fill 87 per cent of the enlisted complement....Without you, we would be seriously handicapped."(270)

because it was doubtless representative of the thinking of so many Marines, was made by a young corporal wounded at Guadalcanal, who said:

"Well, I'll tell you. I was kinda sore about it [the Women Marines] at first. Then it began to make sense--though only if the girls are gonna be tops, understand."

His friend, a sergeant, broke in. "Hell," he said, "they're gonna be Marines, aren't they? They gotta be tops!"(271)

NOTES

(1) "Women in the War," 1944 Office of War Information folder (Subject File "WOMEN MARINES--Enlistment," Historical Branch, Headquarters, U. S. Marine Corps, hereafter HistBr, HQMC), pp. 6-7. WAACs, 15May42; WAVES, 30Jul42; SPARS, 23Nov42; Marines, 13Feb43.

(2) Ibid., p. 9.

(3) Typed excerpt Chap XXXI "Women in the Navy," from Josephus Daniels, Our Navy at War (Subject File "WOMEN MARINES--World War I," HistBr, HQMC) pp. 328-329

(4) Ibid.

(5) Col Ruth Cheney Streeter and LtCol Katherine A. Towle, "History of the Marine Corps Women's Reserve - A Critical Analysis of Its Development and Operation, 1943-1945," dtd 6Dec45 (HistBr, HQMC), p. 21, hereafter Streeter, "History."

(6) Commandant of the Marine Corps (hereafter CMC) ltr to Secretary of the Navy (hereafter SecNav), dtd 12Oct42 (Folder "1535-55-10 Female Enrollment MCR #1," Central Correspondence Files, HQMC, hereafter CCF, HQMC).

(7) Ibid., Judge Advocate General 1st End, dtd 26Oct42.

(8) Ibid., Commander-in-Chief U. S. Fleet 2d End, dtd 30Oct42.

(9) Ibid., SecNav approval, dtd 31Oct42 and the President's approval, dtd 7Nov42.

(10) CMC ltr to SecNav, dtd 14Nov42 (Folder "1535-55-10 Female Enrollment MCR #1," CCF, HQMC).

(11) Col L. W. T. Waller, Jr., memo to CMC, dtd 12Jan43; CMC ltr to SecNav dtd 14Nov42, citing Public Law 689, 77th Congress (Folder 1535-55-10 Female Enrollment MCR #1," CCF, HQMC).

(12) CMC ltr to Commanding Officers of every post and district, dtd 5Nov42 (Folder "1535-55-10 Female Enrollment MCR #1," CCF, HQMC).

(13) Typed carbon draft "Brief of the History of Marine Corps Women's Reserve," Hereafter "Brief History" (Subject File "WOMEN MARINES--History," HistBr, HQMC), p. 3; Streeter, "History," p. 56.

Sources for the Introduction, p.1, follow Note (271)

(14) Division of Reserve (hereafter DivRes) memo to Quartermaster (hereafter QM), dtd 5Jan43 (Folder "1535-55-10 Female Enrollment MCR #1", CCF, HQMC).

(15) CMC ltr to Dean V. C. Gildersleeve, dtd 17Nov42 (Folder "1535-55-10 Female Enrollment MCR #1," CCf, HQMC).

(16) CMC ltr to Dean V. C. Gildersleeve, dtd 3Dec42, (Folder "1534-55-10 Female Enrollment MCR #1," CCF, HQMC); BGen L. W. T. Waller, Jr. typewritten draft (hereafter Waller draft), of Chapter II, Streeter "History," p. 3, with ltr to Col R.C. Streeter, dtd 5Sep45, (Subject File "WOMEN MARINES--History," HistBr, HQMC); Major Carroll B. Rhoads ltr to Mr. Basil O'Connor, Dtd 12Dec42, (Folder "1535-55-10 Female Enrollment MCR #1," CCF, HQMC); USMC release, n.d. (Subject File "WOMEN MARINES--History," HistBr, HQMC).

(17) Waller draft, p. 2; CMC memo to AsstSecNav, dtd 17Dec42, (Folder "1535-55-10 Female Enrollment MCR #1," CCF, HQMC).

(18) Chief of Naval Personnel, CMC, and Commandant Coast Guard joint ltr to SecNav, dtd 20Nov42 (Folder "1535-55-10 Female Enrollment MCR #1," CCF, HQMC; Letter of Instruction (hereafter LI) 281 dtd 11Dec42 (HistBr, HQMC).

(19) Streeter, "History," pp. 22-23; Women Marines (hereafter WM) Director's Office telephone conversation with HistBr, Aug 63.

(20) Col Waller memo to CMC, dtd 12Jan43 (Folder 1535-55-10 Female Enrollment MCR #1," CCF, HQMC); The name SPAR was taken from the Coast Guard motto, "Semper Paratus - Always Ready".

(21) Washington, D. C. Times-Herald, dtd 16Feb43 (Subject File "WOMEN MARINES--News Clippings #1," HistBr, HQMC).

(22) Washington, D, C. Star, dtd 15Feb43 (Subject File WOMEN MARINES--Officer Training," HistBr, HQMC).

(23) San Diego Marine Corps Base Chevron, dtd 5Aug44 (Subject File "WOMEN MARINES--News Clippings #1," HistBr, HQMC).

(24) LI 281, dtd 11Dec42; "Brief History," p. 4.

(25) Ibid.

(26) Ibid.

(27) Streeter, "History," pp. 23-24.

(28) Officer-in-Charge (hereafter OIC), San Francisco, Calif. Procurement Office ltr to BGen L. W. T. Waller, Jr., dtd 26Feb43, (Folder "1535-55-10 Female Enrollment MCR #2," CCF, HQMC).

(29) Miss G. I. Filson ltr, n.d., to MajR.C. Streeter (Folder "1535-55-10 Female Enrollment MCR #4," CCF, HQMC) Avonmore, Pa. girl ltr, to Mrs. F. D. Roosevelt, dtd 31Mar43 (Folder "1535-55-10 Female Enrollment MCR #2," CCF, HQMC).

(30) Mr. J. Carter ltr to CMC, dtd 31Jan43, (Folder "1965-90-10-5 FEMALE Appointment #1," CCF, HQMC).

(31) Col W. W. Rogers, M-3, memo to Director of Plans and Policies (hereafter Dir, DivP&P), dtd 6Mar43; Greater New York Federation of Chruches to HQMC and other correspondence, dtd 15Feb43 (Folder "1535-55-10 Female Enrollment MCR #2," CCF, HQMC).

(32) Col L. W. T. Waller, Jr., ltr to Hon. L. Ludlow, dtd 8Feb43 (Folder "1535-55-10 FEMALE Enrollment MCR #1," CCF, HQMC).

(33) CMC ltr SecNav, dtd 18Dec42, (Folder "1535-55-10 FEMALE Enrollment MCR #1," CCF, HQMC).

(34) Procurement Directive No. 17, dtd 10Mar43 (Folder "1535-55-10 Female Enrollment MCR #2," CCF, HQMC).

(35) Navy Dept Release, dtd 17Mar43 (Subject File "WOMEN MARINES--Enlistment," HistBr, HQMC).

(36) Ibid.; "Itinerary Maj Streeter's trip of 17Feb-26Mar43," (Subject File "Women's Reserve--USMCR," HistBr, HQMC).

(37) Maj C. B. Rhoads ltr to Mr. Thomas Streeter, dtd 8Mar43 (Folder "1965-90-10-5 FEMALE Appointments #1," CCF, HQMC).

(38) Navy Dept release, dtd 9Apr43 (Subject File "WOMEN MARINES--Enlistment," HistBr, HQMC).

(39) Ibid.; Release, dtd 15Apr43 (Subject File "WOMEN MARINES--History," HistBr, HQMC).

40) OIC Chicago, Ill. Procurement Office ltr. to LtCol John R. Moe, dtd 15Mar43 (Folder "1535-55-10 Female Enrollment MCR #2," CCF, HQMC).

(41) Streeter, "History," p. 24; Speed ltr signed Waller & Lawton to Directors & OICs, dtd 18Mar43 (Folder "1535-55-10 Female Enrollment MCR #2," CCF, HQMC); CMC ltr to SecNav, dtd 24Jul 1943 (Folder 1535-55-10 Female Enrollment MCR #3,' CCF, HQMC).

(42) Release, dtd 6Nov43 (Subject File "WOMEN MARINES--Enlistment," HistBr, HQMC); Marine Corps release, "Anniversary Message to Marine Corps Women's Reserve" from Col R. C. Streeter (Subject File "WOMEN MARINES--Anniversaries," HistBr, HQMC).

(43) Director R. C. Streeter memo to Acting Director of Personnel, dtd 24May44 (Folder "1535-55-10 Female Vol 1-P," CCF, HQMC).

(44) Streeter, "History," pp. 26, 280; Navy Dept release, dtd 12Mar43 (Subject File "WOMEN MARINES--Officer Training," HistBr, HQMC).

(45) Ibid.

(46) MajGen H. Schmidt memo to Chief of Naval Personnel (hereafter Schmidt memo), dtd 23Feb43 (Folder "1535-55-10 Female Enrollment MCR #2," CCF, HQMC).

(47) CMC ltr to SecNav, dtd 6Jul43, refers to SecNav approval of "Cadet" rank on 26Dec43; Release, dtd 15Apr43 (Subject File "WOMEN MARINES--History," HistBr, HQMC); LI 382, dtd 27Mar43; One member of the second class recalled that candidates wore arm bands rather than OC pins throughout training. Information WM Director's Office, Feb64 (Monograph and Comment File, HistBr, HQMC)

(48) Streeter, "History," p. 280.

(49) Schmidt memo.

(50) LI 382, dtd 27Mar43.

(51) Schmidt memo; Streeter, "History," p. 25; Navy Dept release, dtd 21Mar43 (Subject File "WOMEN MARINES--Recruit Training, "HistBr, HQMC); Col K. A. Towle, "Women Marines: The Feminine Side," Marine Corps Gazette, v. 34, no. 11 (Nov50), p. 111, hereafter Towle, "Women Marines."

(52) Schmidt memo; Col L. W. T. Waller, Jr., memo to Col B. W. Gally, dtd 6Mar43 (Folder "1535-55-10 Female Enrollment MCR #2," CCF, HQMC).

(53) Ibid.; "Brief History," p. 1; Navy Dept release, dtd 21Mar43 (Subject File "WOMEN MARINES--Recruit Training," HistBr, HQMC); WM Director's Office telephone coversation with HistBr, Aug63.

(54) Navy Dept release, dtd 21Mar43 (Subject File "WOMEN MARINES-Recruit training," HistBr, HQMC).

(55) Streeter, "History," pp. 25, 117 (attrition figure computed from figures on p. 117).

(56) HQMC memo, dtd 3Mar43 (Folder "1535-55-10 Female Enrollment MCR #2," CCF, HQMC).

(57) CMC memo to Chief of Naval Personnel, dtd 10Apr43 (Folder "1965-90-10-5 FEMALE Appointments #1, CCF, HQMC); Towle, "Women Marines," p. 111; Director R. C. Streeter memo to Director, Personnel Dept, dtd 5Jul43 (Folder "1965-90-10-5 Female Appointments #2, CCF, HQMC).

(58) MajGen H. Schmidt ltr to Dr. R. G. Ham, dtd 23Jul43 (Folder "1535-55-10 Female Enrollment MCR #2," CCF, HQMC).

(59) MajC. B. Rhoads memo to Director, Personnel Dept, dtd 24Jun43 (Folder "1965-9--10-5 Female Appointments #2," CCF, HQMC).

(60) CMC memo to Chief Naval Personnel, dtd 24May43 and HQMC correspondence, dtd 15May43 (Folder "1535-55-10 Female Enrollment MCR #3," CCF, HQMC).

(61) Waller draft, p. 3.

(62) Streeter "History," p. 123.

(63) Ibid.

(64) Extract from personal ltr received by BGen L. W. T. Waller, Jr., from Maj E. H. Hurst, forwarded as part of an official memo from BGen Waller to BGen K. E. Rockey, dtd 3Apr43 (Folder "1535-55-10 Female Enrollment MCR #2," CCF, HQMC).

(65) Director R. C. Streeter memo to Director, Personnel Dept, dtd 12Jun43 (Folder "1965-90-10-5 Female Appointments #2," CCF, HQMC).

(66) Radio script, dtd 14Jan44, Ser #164024 (Subject File "WOMEN MARINES--Enlistment," HistBr, HQMC, p. 3.

(67) Marine Corps release, "Training Women Reserve at Camp Lejeune, New River, N. C.," n.d.; Release carbon draft pp. 8, 10 (Subject File "WOMEN MARINES--Recruit Training," HistBr, HQMC).

(68) "Broadway Gazette" column of Leonard Lyons, guest-columnist Maj R. C. Streeter, n.d., n.p., (Subject File "WOMEN MARINES--History," HistBr, HQMC), hereafter Broadway column; Robert H. Myers, "Boot Camp for Women - Part I," Leatherneck, Vol. 26, no. 9 (Sep43), p. 44.

(69) Guy Richards, "The Ladies Arrive--The Story of the Women's Reserve, (a monograph in posession of Col W. P. McCahill, USMCR, Washington, D. C.

(70) 1stLt J. A. Kelly, Atlanta, Ga. Procurement Office ltr to HQMC, dtd 27Mar43 (Folder "1535-55-10 Female Enrollment MCR #2," CCF, HQMC).

(71) HQMC memo, dtd 13Aug43 (Folder "1535-55-10 Female Enrollment MCR #4," CCF, HQMC); Director R. C. Streeter telegram to Philadelphia, Pa. Platoon dtd 4Sep43 (Folder "1535-55-10 Female Enrollment CCF #3, HQMC).

(72) Lt Virginia L. McCance, Procurement Office, Pittsburgh, Pa. ltr to HQMC, dtd 29Nov43 (Folder "2295-100 Women's Reserve," CCF, HQMC); Luther H. Evans, Chief Assistant Librarian, Library of Congress ltr, to Lt Frances M Seibert, Officer Procurement District, Washington, D. C., dtd 18Nov43 (Folder "1535-55-10 Female Enrollment MCR #4," CCF, HQMC).

(73) HQMC ltrs, dtd 9Oct43 and 18Oct43 (Folder "1535-55-10 Female Enrollment MCR #4," CCF, HQMC); HQMC memo, dtd 13Aug43, op. cit.; Correspondence, dtd 19Jan44 (Folder "1535-55-10 Female Vol. 1-P," CCF, HQMC).

(74) "Brief History," p. 5; Streeter, "History," p. 118; Waller draft, p. 3.

(75) Marine Corps release, "Training Women Reserves at Camp Lejeune, New River, N. C.," op. cit.; HQMC memo, dtd 13Aug43, op.cit.; "Brief History," p. 5.

(76) Streeter, "History," p. 126; "Brief History," p. 5.

(77) Dir, DivRes ltr to Dir, Training Division, Bureau of Naval Personnel, dtd 19Apr43 (Folder "1535-55-10 Female Enrollment MCR #3," CCF, HQMC).

(78) Marine Corps Women's Reserve (hereafter MCWR) release, n.d., (Subject File "WOMEN MARINES--Section Reports," HistBr, HQMC); Streeter, "History," pp. 129-130.

(79) Radio script, dtd 14Jan44, op. cit., p. 4

(80) MCWR release, n.d. (Subject File "WOMEN MARINES--Section Reports," HistBr, HQMC); LI 574, dtd 30Oct43.

(81) "Brief History," pp. 7-8.

(82) CMC ltr to OICs of four Procurement Districts, dtd 27Jul43 (Folder "1535-55-10 Female Enrollment MCR #3," CCF, HQMC); LI, dtd 6Jul43; LI, dtd 3May44; Streeter, "History," p. 256.

HQMC correspondence, dtd 3Sep43 (Folder "1965-90-10-5 Female Appointments #2," CCF, HQMC).

(83) Marine Corps release, dtd 22Jul43 (Subject File "WOMEN MARINES--Officer Training," HistBr, HQMC); CMC ltr to CO, MCWR Schools, dtd 30Jul43 (Folder "1965-90-10-5 Female Appointments #2," CCF, HQMC); Draft Chap 14, "MCWR Officer Training," pt 3 (Folder "2185-65-10 History of MCWR Supplementary, Vol. I," CCF, HQMC).

(84) LI 482, dtd 6Jul43; LI 573, dtd 30Oct43; LI 739, dtd 3May44; CMC correspondence to SecNav, dtd 6Jul43 (Folder "1965-90-10-5 FEMALE Appointments #2," CCF, HQMC).

(85) Capt F. L. Churchville memo to Dir, DivRes, dtd 7May43 (Folder "1965-90-10-5 FEMALE Appointments #1," CCF, HQMC); HQMC correspondence, dtd 24Jun43 and 3Sep43 (Folder "1965-90-10-5 Female Appointments #2," CCF, HQMC); Dir, P&P memo to CMC dtd 17Mar45 (Folder "1965-90-10-5 Appointments Female-1," CCF, HQMC).

(86) Director R. C. Streeter memo to Dir, Personnel Dept, dtd 18Dec44 (Folder "1965-90-10-5 Appointments Female-1-," CCF, HQMC); Streeter, "History," pp. 284-287.

(87) Streeter, "History," p. 64; CMC memo to AsstSecNav, dtd 17Dec42 (Folder "1535-55-10 FEMALE Enrollment MCR #1," CCF, HQMC).

(88) Navy Dept release, dtd 20Mar43 (Subject File "WOMEN MARINES--Uniforms," HistBr, HQMC); Col L. W. T. Waller, Jr., memo to Mrs. A. A. Lentz, dtd 31Dec42 (Folder "1535-55-10 FEMALE Enrollment MCR #1," CCF, HQMC).

(89) NavyDept release, dtd 20Mar43, op. cit.

(90) Waller draft, p. 3.

(91) Ibid., p. 2; Correspondence, dtd Feb43 with Harper's Bazaar clipping (Folder "1535-55-10-5 Female Enrollment MCR #2," CCF, HQMC).

(92) Mr. E. A. Greene ltr to Director R. C. Streeter, dtd 6Feb43 (Folder "1965-90-10-5 FEMALE Appointments #1," CCF, HQMC).

(93) The Institute for Research, "A Career in the U. S. Marine Corps Women's Reserve," Chicago, 1943, hereafter Career booklet (Subject File "WOMEN MARINES--Enlistment," HistBr, HQMC).

(94) Adjutant & Inspector ltr to Dir, DivRes, dtd 21Nov42; Col L. W. T. Waller, Jr., memo to CMC, dtd 12Jan43 (Folder "1535-55-10 FEMALE Enrollment MCR #1," CCF, HQMC); HQMC advance information bulletin, dtd 6Feb43 (Subject File "WOMEN MARINES--History," HistBr, HQMC); LI 424 dtd 15May43; Dir, DivRes ltr to Research Section, dtd 5Dec42 (Folder "1535-55-10 Female Enrollment MCR #1," CCF, HQMC); Myers, "Boot Camp for Women," *op. cit.*, p. 44.

(95) Career Booklet, p.7; Col R.H. Rankin, USMC, *Uniforms of the Sea Services--A Pictorial History* (Annapolis: U. S. Naval Institute, 1962), pp. 252-253; USMC release, n.d. (Subject File "WOMEN MARINES--History," HistBr, HQMC).

(96) *Waller draft*, p. 3; Navy Dept release, dtd 20Mar43 (Subject File "WOMEN MARINES--Uniforms," HistBr, HQMC); LtCol Ruth H. Broe ltr to Head, HistBr, G-3 Div, HQMC, dtd 15Nov63 (Monograph and Comment File, HistBr, HQMC).

(97) Rankin, *Uniforms*, *op. cit.*, pp. 252-253; Col Margaret M. Henderson ltr to Assistant Chief of Staff, G-3, dtd 14Nov63 (Monograph and Comment File, HistBr, HQMC).

(98) Rankin, *Uniforms*, *op. cit.*, p. 253

(99) Uniform Regulations, 1943, USMCR, revised 27Jul43 and issued as LI 523, dtd 27Aug43; LI 737, dtd 1May44

(100) LI 523, *op. cit.*

(101) *Ibid.*

(102) *Waller draft*, p. 3; LI 602, dtd 27Nov42.

(103) Typed draft of release, dtd 9Sep43, and Bob Jackson, Philadelphia, Pa. Procurement Office memo to S/Sgt MacQueen, of *Recruiter* (Subject File "Women's Reserve - USMCR," HistBr, HQMC).

(104) "Brief History," p. 3; Streeter, "History," pp. 64-65; Typed carbon of release, "First Class of Uniform Officers," dtd 15Nov43 (Subject File "WOMEN MARINES--Officer Training," HistBr, HQMC); Release #40, "Women's Post Exchange at Camp Lejeune," n.d. (Subject File "WOMEN MARINES--Recruit Training," HistBr, HQMC).

(105) "Brief History," p. 3; Streeter, "History," pp. 64, 66, 70.

(106) Streeter, "History," p. 69.

(107) *Ibid.*, pp. 73-74.

(108) Capt F. R. Washington, Asst to Dir, DivRes ltr to Mr. F.M. Schlegel, dtd 2Mar43, (Folder "1965-90-10-5 FEMALE Appointments #1," CCF, HQMC); Broadway column.

(109) Typed excerpt Chap XXXI, "Women in the Navy," from Josephus Daniels, Our Navy at War, op. cit.

(110) Towle, "Women Marines," p. 111; Lt. E. Louise Stewart, "What Are the Women Marines Doing?" Marine Corps Gazette, v. 27, no. 6 (Oct43) passim.

(111) Headquarters Bulletin, dtd Oct43 (Subject File "WOMEN MARINES--News Clippings #1," HistBr, HQMC); "Brief History," addendum pp. 1-5.

(112) Navy Dept release, dtd 23Mar43 (Subject File "WOMEN MARINES--Section Reports," HistBr, HQMC).

(113) Ibid.

(114) "Brief History," p. 5.

(115) Ibid., MCWR release, n.d. (Subject File "WOMEN MARINES--Section Reports," HistBr, HQMC).

(116) Streeter, "History," pp. 159, 162-164.

(117) Ibid., p. 161.

(118) Ibid., p. 204; "Brief History," p. 6; LI 574, dtd 30Oct43.

(119) "Brief History," p. 9; LI 622, dtd 29Dec43; Director R. C. Streeter memo to CMC, dtd 22Dec43 (Folder: "1965-90-10-5 Female Appointments #2," CCF, HQMC).

(120) Waller ltr to Mr. Fields, dtd 11May43 (Folder "1965-90-10-5 FEMALE Appointments #1," CCF, HQMC); "Women Reserves Celebrate their Second Anniversary," Chevron, 10Feb45 (Subject File "WOMEN MARINES--Anniversaries," HistBr, HQMC), hereafter Chevron article.

(121) Ibid., Stewart, "What Are Women Marines Doing?" op. cit., p. 30.

(122) Chevron article.

(123) Quantico Sentry, dtd Nov43 (Subject File "WOMEN MARINES--News Clippings #1," HistBr, HQMC).

(124) Chevron article.

(125) Ibid.

(126) Ibid.

(127) Ibid.

(128) Ibid.

(129) Ibid.

(130) Robert Sherrod, History of Marine Corps Aviation in World War II (Washington: Combat Forces Press, 1952), p. 128; Streeter, "History," pp. 127-128.

(131) Ibid.; Division Aviation memo to Division Reserve, dtd 25Feb43 (Folder "1965-90-10-10-5 FEMALE Appointments #1," CCF, HQMC).

(132) HQMC memo, dtd Sep43 (Folder "1965-90-10-5 Female Appointments #1," CCF, HQMC).

(133) Carbon draft of release, dtd 2Aug44 (Subject File, "WOMEN MARINES--Specialist Training," HistBr, HQMC).

(134) Draft of LI 1093, dtd 4Aug45 (Folder "2295-100 Report on Women's Reserve," CCF, HQMC); Muster Rolls, Diary Unit, Files Section, Personnel Department, HQMC; AWRS - 12 and - 13 at Edenton were disbanded early in 1945.

(135) Streeter, "History," pp. 165-168.

(136) Director R. C. Streeter memo to All Officers, dtd 17Jul43 (Folder "1535-55-10 Female Enrollment MCR #3," CCF, HQMC).

(137) John A. De Chant, Devilbirds (New York: Harper & Bros., 1947), pp. 58-59.

(138) Streeter, "History," pp. 375-376.

(139) Washington Star, 13Sep45 (Subject File "WOMEN MARINES--News Clippings #1," HistBr, HQMC).

(140) "Brief History," p. 2; Streeter, "History," pp. 21-22, 45.

(141) Streeter, "History," p. 31.

(142) LI 382, dtd 27Mar43.

(143) LI 571 dtd 29Oct43; Streeter, "History," p. 32.

(144) Waller draft, p. 4; Release, dtd 25Nov43, CMC to All Officers, MCWR (Subject File "WOMEN MARINES--Regulations," HistBr, HQMC); Streeter, "History," p. 9; LtCol John B. Hill report to Dir, DivRes, dtd 4Feb43 (Folder "2295-100 Women's Reserve," CCF, HQMC).

(145) Correspondence, dtd 18Mar43 (Folder "1535-55-10 Female Enrollment MCR #2," CCF, HQMC); Correspondence, dtd 21May43 and 31May43 (Folder "1535-55-10 Female Enrollment MCR #3,: CCF, HQMC); Correspondence, dtd 22Nov43 (Folder "1535-55-10 Female Enrollment MCR #4," CCF, HQMC); Mattie E. Treadwell, U. S. Army in World War II--Special Studies--The Women's Army Corps (Washington: Office of the Chief of Military History, Dept of the Army, 1954), pp. 483-484, 759.

(146) Correspondence, dtd 31May43 (Folder "1535-55-10 Female Enrollment MCR #3," CCF. HQMC).

(147) Carbon of release, n.d. (Subject File "WOMEN MARINES--Officer Training," HistBr, HQMC).

(148) CMC ltr, dtd 20Jul43 (Folder "1535-55-10 Female Enrollment MCR #3," CCF, HQMC).

(149) Ibid.

(150) Director R. C. Streeter memo to All Officers, dtd 17Jul43 (Folder "1535-55-10 Female Enrollment, MCR #3," CCF, HQMC).

(151) CMC ltr, dtd 25Nov43, op. cit.; CMC ltr, dtd 20Jul43, op. cit.; BGen K. E. Rockey memo to DivRes, dtd 8Mar43 (Folder "1535-55-10 Female Enrollment MCR #2," CCF, HQMC).

(152) Streeter, "History," pp. 384-385; Commanding Officer, 2d Headquarters Battalion memo, dtd 26Jul43, Subj: Gripe Sheet (Subject File "WOMEN MARINES--Section Reports," HistBr,HQMC), p. 10.

(153) LI 725, dtd 25Apr44; Streeter, "History," pp. 363-364.

(154) LI 571, dtd 29Oct43.

(155) Ibid.; LI 726, dtd 26Apr 44.

(156) Director R. C. Streeter ltr to James M. Nelson, Assoc Ed., The American Magazine, dtd 28Sep43 (Folder 1535-55-10 Female Enrollment MCR #4," CCF, HQMC).

(157) LI 382, dtd 27Mar43.

(158) CMC ltr, dtd 25Nov43, op. cit.

(159) LI 692, dtd 27Mar44 which refers to LI 382, dtd 27Mar43.

(160) Ibid.

(161) LI 291, dtd 11Dec42.

(162) SecNav approval of modification of marriage regulations, dtd 9Mar43 (Folder "1535-55-10 Female Enrollment MCR #2," CCF, HQMC).

(163) Director R. C. Streeter memo to Procurement Branch, dtd 10Aug43 (Folder "1535-55-10 Female Enrollment MCR #4," CCF, HQMC).

(164) Ibid.

(165) Policy change statements, dtd 17Nov43 and 26Nov43 (Folder "1535-55-10 Female Enrollment MCR #4," CCF, HQMC).

(166) LI 382, dtd 27Mar43; LI 510, dtd 6Aug43; LI 499, dtd 29Jul43; LI 858, dtd 5Oct44; Streeter, "History," p. 14; "Brief History," pp. 6-7.

(167) Ibid.; Streeter, "History," pp. 43-44.

(168) Director R. C. Streeter memo, dtd 17Jul43, op. cit.

(169) Ibid.

(170) Broadway column; Navy Dept release, dtd 10Feb44 (Subject File "WOMEN MARINES--Officer Training," HistBr, HQMC); Capt F. L. Churchville memo to Dir, DivRes, dtd 7Apr43 (Folder "1965-90-10-5 FEMALE Appointments #1," CCF, HQMC); muster rolls, Diary Unit, Files Section, Personnel Department, HQMC.

(171) 1stLt E. Louise Stewart, "Shakedown Cruise," Marine Corps Gazette, vol. 27, no. 2 (May-June43), p. 37.

(172) USMC release, dtd 15Apr43 (Subject File "WOMEN MARINES--History," HistBr, HQMC); Navy Department release, dtd 4Feb44 (Subject File "WOMEN MARINES--Officer Training," HistBr, HQMC).

(173) Correspondence, dtd 24Jun43 (Folder "1965-90-10-5 Female Appointments #2," CCF, HQMC); #34, n.d. (Subject File "WOMEN MARINES--Press Releases," HistBr, HQMC).

(174) USMC release, dtd 1Feb45 (Subject File "WOMEN MARINES--Anniversaries," HistBr, HQMC), pp. 2-3; Disabled American Veterans Semi-Monthly, dtd Jan44 (Subject File "WOMEN MARINES--News Clippings #1," HistBr, HQMC).

(175) Ibid., pp. 1-2; Carbon of Navy Dept release, dtd Mar43 (Subject File "WOMEN MARINES--Press Releases," HistBr HQMC); Navy Dept release, dtd 10Feb44 (Subject File "WOMEN MARINES--Officer Training," HistBr, HQMC).

(176) BGen L. W. T. Waller, Jr., memo, dtd 13Jul43 (Folder "1535-55-10 Female Vol. 1-P," CCF, HQMC).

(177) BGen L. W. T. Waller, Jr., ltrs to various Directors, Musical Schools, dtd 31Jul43 (Folder "1535-55-10 Female Vol. 1-P," CCF, HQMC); Streeter, "History," pp. 376-377; Carbon of USMC release, New River, N. C., dtd 11Nov45 (Subject File "WOMEN MARINES--Specialist Training," Hist-Br, HQMC).

(178) Streeter, "History," p. 377; Towle, "Women Marines," p. 112.

(179) San Diego *Chevron*, dtd 10Feb45 (Subject File "WOMEN MARINES--Anniversaries," HistBr, HQ C).

(180) Sgt Chas B. Kopp, USMC typed release, nd (Subject File "WOMEN MARINES--Press Releases," HistBr, HQMC).

(181) Navy Dept release, dtd 2Aug43 (Subject File "WOMEN MARINES--Press Releases," HistBr, HQMC); Carbon of Camp Lejeune, N. C. release, dtd Sep43 (Subject File "WOMEN MARINES--Recruit Training," HistBr, HQMC).

(182) *Ibid*.

(183) Carbon of release, Camp Lejeune, N. C., dtd 6Dec43 (Subject File "WOMEN MARINES--Press Releases," HistBr HQMC); Washington *Times-Herald*, dtd 23Nov44 (Subject File "WOMEN MARINES--News Clippings #1," HistBr, HQMC).

(184) USMC release, dtd 24May44 (Subject File "WOMEN MARINES--Press Releases," HistBr, HQMC); Henderson ltr, *op. cit*.

(185) Navy Dept release, dtd 6Aug43 (Subject File "WOMEN MARINES--Press Releases," HistBr, HQMC); *Headquarters Bulletin*, dtd Oct43 (Subject File "WOMEN MARINES--News Clippings #1," Histbr, HQMC); HQMC correspondence, dtd 3Jun43 (Folder 1535-55-10 Female Vol. 1-P," CCF, HQMC); Carbon release, dtd 12 Apr 44 (Subject File "WOMEN MARINES--Press Releases," HistBr, HQMC).

(186) Washington *Star* picture, dtd 6Feb44 (Subject File "WOMEN MARINES--News Clippings #1," HistBr, HQMC).

(187) Carbon of USMC release, dtd 17Nov43 (Subject File "WOMEN MARINES--Officer Training," HistBr, HQMC); Mrs. F. A. Bunte ltr to Director R. C. Streeter, dtd 24Jul43 (Folder "1535-55-10 Female Vol. 1-P," CCF, HQMC).

(188) USMC release, dtd 13Feb45 (Subject File "WOMEN MARINES--Anniversaries," HistBr, HQMC); Washington *Times-Herald* articles, dtd 13Apr45 and 28Aug45 (Subject File "WOMEN MARINES--News clippings #1," HistBr, HQMC).

(189) San Diego *Chevron*, dtd 10Feb45 (Subject File "WOMEN MARINES--Anniversaries," HistBr, HQMC).

(190) Typed draft, dtd 7Oct43 (Subject File "WOMEN MARINES--Press Releases," HistBr, HQMC); USMC release, dtd 1Feb45 (Subject File "WOMEN MARINES--Anniversaries," HistBr, HQMC).

(191) HQMC release, dtd 1Feb45 (Subject File "WOMEN MARINES--Anniversaries," HistBr, HQMC). passim.

(192) Head, Decorations and Medals Branch memo to HistBr, HQMC, dtd 5Sep63.

(193) Director R. C. Streeter memo to Acting Director of Personnel, dtd 13Sep44 (Folder "1535-55-10 Female Vol. 1-P," CCF, HQMC); HQMC Legal division telephone conversation with HistBr, Jul63; 11th Naval District Bulletin, Serial No. LA(a) 188(43) (Subject File "WOMEN MARINES--Overseas Duty," HistBr, HQMC).

(194) Telephone conversation, Jul63, op. cit.; Streeter memo, dtd 13Sep44, op. cit.

(195) Washington Times-Herald, 20Sep44 (Subject File "WOMEN MARINES--News Clippings #1," HistBr, HQMC).

(196) Telephone conversation, Jul63, op. cit.

(197) Streeter memo, dtd 13Sep44, op. cit., refers to July Hawaii memorandum.

(198) Director R. C. Streeter memo, dtd 3Nov44, Subj: Report of Visit to Hawaii 13Oct-2Nov (Folder "2295-100 Report on Women's Reserve, 1Jan44-31Dec47," CCF, HQMC); Streeter, "History," pp. 387-388.

(199) CMC memo to WR Assistants, dtd 30Jan45 (Folder "1535-55-10 Female Vol. 1-P," CCF, HQMC); Release draft "Wahine Marine," AO3E-1265-HPH, dtd 22Jan53 (Subject File "WOMEN MARINES--History,: HistBr, HQMC).

(200) LI 884, dtd 13Nov44.

(201) Ibid.; BGen Waller memo to Director R. C. Streeter, dtd 30Oct43 (Folder "1535-55-10 Female Vol. 1-P" CCF, HQMC).

(202) HQMC correspondence (Folder "1535-55-10 Female Vol. 1-P" CCF, HQMC).

(203) Streeter memo, dtd 3Mov44, op. cit.; USMC release "WM Officers on Duty in Hawaii,:n.d. (Subject File "WOMEN MARINES--Press Releases," HistBr, HQMC); Information WM Director's Office to HistBr, Jun63; LtCol D. A. Stafford memo to CMC re rank of officers, dtd 15Feb43 (Folder "1965-90-10-5 FEMALE Appointments #1," CCF, HQMC).

(204) Information WM Director's Office, Jun63; Release "Headquarters Announces Qualifications for Women Marines Going Overseas," dtd Nov44 (Subject File "WOMEN MARINES--Overseas," HistBr, HQMC)

(205) Information WM Director's Office, Jun63; LtCol Mary Hale conversation with HistBr, Aug63.

(206) HQMC correspondence, dtd 2Jan45 (Folder "1535-55-10 Female Vol. 1-P," CCF, HQMC).

(207) Marine Corps Women's Reserve Battalion paper, v.1, no. 1 (May45) (Subject File "WOMEN MARINES--Overseas," HistBr, HQMC); Wahine Marine" release, op. cit.

(208) Ibid.; Director R. C. Streeter memo (Folder "1535-55-10 Female Vol. 1-P" CCF, HQMC); Information WM Director's Office, Jun63.

(209) "Wahine Marines" release, op. cit.

(210) Washington Times-Herald, 18Dec44 (Subject File "WOMEN MARINES--News Clippings #1," HistBr, HQMC).

(211) Reserve battalion paper, op. cit.

(212) Ibid.

(213) Typed draft "Women Marines" #145-WR Motor Transport, dtd May45 (Subject File "WOMEN MARINES--Press Release," HistBr, HQMC).

(214) Sgt Jean Kautenberg, draft of release #148-WR, dtd 30May45 (Subject File "WOMEN MARINES--Press Releases," HistBr, HQMC).

(215) Information WM Director's Office, Jun63; Reserve battalion paper, op. cit.

(216) "The Ladies Arrive," op. cit.

(217) USMC release, n. d., with dateline "Oahu, T.H. (Delayed)" (Subject File "WOMEN MARINES--Press Releases," HistBr, HQMC); Streeter, "History," p. 9.

(218) Ibid., p. 390; Towle, "Women Marines," p. 112.

(219) Release "The Marine Corps Women's Reserve," n. d. (Subject File "WOMEN MARINES--Reserve Section Reports" HistBr, HQMC).

(220) Streeter, "History," p. 419; Draft "Marine Corps Women's Reserve," op. cit.

(221) HQMC Bulletin 1355-40 DDA-483 hdk, dtd 16Oct45 (Subject File: "WOMEN MARINES--History," HistBr, HQMC).

(222) *Ibid*.

(223) Streeter, "History," pp. 378-379; Aviation Women's Reserve Squadron (AWRS) 12 War Diary and History, Apr44-Dec44 (HistBr, HQMC), p. 27; Correspondence dtd 9Jan46 (Folder "2295-100 Report of Women Reserve (T)," CCF, HQMC).

(224) Streeter, "History," pp. 420-421; LI 1110, dtd 1Aug45.

(225) Streeter, "History," p. 424.

(226) Washington *Post*, 2Nov45 (Subject File "WOMEN MARINES--News Clippings #1," HistBr, HQMC).

(227) Biographies of the directors (Subject File "WOMEN MARINES," HistBr, HQMC).

(228) Lt Mary L. Parks monthly report to HQMC, dtd 4Jan45 (Folder "2295-100 Report of Women Reserve (T)," CCF, HQMC).

(229) Streeter, "History," p. 100.

(230) Draft of LI, dtd 4Aug45 (Folder "2295-100 Report of Women's Reserve 1Jan44-31Dec47," CCF, HQMC).

(231) *Ibid*.

(232) *Ibid*.

(233) *Ibid*.

(234) Streeter, "History," pp. 420-421.

(235) *Ibid*.

(236) AWRS 3 monthly report, dtd 6Dec45 (Folder "2295-100 Report of Women Reserve (T)," CCF, HQMC); Henderson Hall, Arlington, Va. monthly report, dtd 9Jan46 (Folder as above).

(237) AWRS 4 monthly report, dtd 19Dec45 (Folder "2295-100 Report of Women Reserve (T)," CCF, HQMC).

(238) Henderson Hall, Arlington, Virginia monthly report, *op. cit*.

(239) Rehabilitation Section, Women's Reserve Separation Company, Camp Lejeune, N. C. monthly report, dtd 31Dec45 (Folder "2295-100 Report of Women Reserve (T)," CCF, HQMC).

(240) Southeastern Recruiting Division, Atlanta, Ga. monthly report, dtd 4Jan46 (Folder "2295-100 Report of Women Reserve (T)," CCF, HQMC).

(241) LI 1110, dtd 21Aug45; Streeter, "History," p. 426.

(242) Ibid.

(243) USMC release, dtd 30Jun45 (Subject File "WOMEN MARINES--Separation from Service," HistBr, HQMC); Director K. A. Towle report to Director, Personnel, dtd 14Jun46, Subj: date of termination of office of Director of wartime Women's Reserve and her release from duty (Folder "2295-100 Report on Women's Reserve 1Jan44-31Dec47," CCF, HQMC).

(244) Ibid.

(245) Ibid.

(246) Streeter, "History," pp. 16, 270.

(247) Ibid.

(248) Commanding Officer, 2d Headquarters Battalion memo, op. cit., p. 10.

(249) Career Booklet, p. 21; Picture Washington Times Herald, 14Jun43 (Subject File "WOMEN MARINES--News Clippings #1," HistBr, HQMC).

(250) AWRS-12 WD&History, op. cit., pp. 10, 24.

(251) Director R. C. Streeter ltr to Capt Bart Dutto, dtd 29Jul43 (Folder "1535-55-10 Female Enrollment MCR #3," CCF, HQMC).

(252) Release "The Marine Corps Women's Reserve," n. d. (Subject File "WOMEN MARINES--Section Reports," HistBr, HQMC), pp. 4-5.

(253) Ibid.

(254) Ibid.; regional totals projected from Appendix III.

(255) Director R. C. Streeter ltr to Capt Linus N. Hardy, dtd 5Aug43 (Folder "1535-55-10 Female Enrollment MCR #4," CCF, HQMC); OIC Procurement ltr to CMC, dtd 17Dec43. (Folder as above).

(256) CMC ltr to Dean Gildersleeve, dtd 12Jan43 (Folder "1965-90-10-5 Female Appointments #1," CCF, HQMC); Ltr to AssistDir, MarCorps Officer Procurement Office, San Francisco, Calif., dtd 18Mar43 (Folder as above).

(257) Director R. C. Streeter memo to CMC, dtd 22Dec43 (Folder "1965-90-10-5 Female Appointments #2," CCF, HQMC); Streeter ltr to Dean Gildersleeve, dtd 31Dec43 (Folder as above).

(258) Release, "The Marine Corps Women's Reserve," op. cit.

(259) Streeter, "History," pp. 252, 97-98.

(260) Ibid.

(261) Ibid.

(262) "Brief History," p. 3; Streeter, "History," p. 80.

(263) Ibid.; Commanding Officer, 2d Headquarters Battalion memo, op. cit., p. 17.

(264) Streeter, "History," pp. 94-95.

(265) "Women's Reserve Survey," dtd 21Sep44 (Subject File "Women's Reserve--USMCR," HistBr, HQMC), p. 7.

(266) Streeter, "History," pp. 57-58.

(267) Towle, "Women Marines," p. 112.

(268) "Women in the War," op. cit., p. 9.

(269) Ibid.; Carbon draft "Brief History of Marine Corps Women's Reserve in World War II," n. d. (Subject File "WOMEN MARINES--History," HistBr, HQMC), p. 3.

(270) USMC news clipping, "Women Marines Observe Second Anniversary," dtd 13Feb45, n.p. (Subject File "WOMEN MARINES--Anniversaries," HistBr, HQMC).

(271) Stewart, "Shakedown Cruise," op. cit., p. 37.

Sources for the Introduction

(a) Washington Times-Herald, dtd 10Feb45 (Subject File "WOMEN MARINES--News Clippings #1," HistBr, HQMC)

(b) Janet Wolff, "Influences on Today's Women," What Women Buy (New York: McGraw-Hill Book Co., Inc., 1958), p. 8.

(c) Typewritten carbon draft "Service Completed" (Subject File "WOMEN MARINES--History," HistBr, HQMC), p. 1.

(d) F. T. Stolley, "Humor in Uniform" Reader's Digest, Mar55, v. 66: 26.

Appendix A: Jobs in which Women Marines Were Assigned During World War II

Accountant
Addressing or Embossing Machine Operator
Administrative NCO
Armorer, Aircraft
Artist
Auditor
Automobile Serviceman
Automotive Carburetor & Ignition Mechanic
Automotive Equipment Operator
Automotive Mechanic
Aviation Salvage Crew Mechanic
Aviation Supply Man
Baker
Band Leader
Bandsman, Bass Drum
Bandsman, Clarinet
Bandsman, Cornet or Trumpet
Bandsman, Euphonium or Baritone
Bandsman, Flute or Piccolo
Bandsman, French Horn
Bandsman, Saxophone
Bandsman, Oboe
Bandsman, Snare Drum
Bandsman, Trombone
Bandsman, Tuba
Barracks NCO
Beauty Operator
Boiler Firemen
Bookkeeper
Bookkeeping Machine Operator
Carburetor Mechanic, Aircraft (Designated Type)
Carpenter, Aircraft
Carpenter, General
Cashier
Chaplain's Assistant
Chauffeur
Chemical Laboratory Technician
Chemical Warfare Specialist
Chief Clerk
Chief Ordnance Man, Light Air Fire Control
Classification Specialist
Clearance Desk Clerk
Clerk, Administrative
Clerk, General
Clerk Typist
Code Clerk
Combat Correspondent
Commissary Man
Communication Chief
Control Tower Operator
Cook
Court Reporter
Crane Operator
Dispatcher, Motor Vehicle
Draftsman, Electrical
Draftsman, General
Draftsman, Mechanical
Draftsman, Topographic
Drill Instructor
Drum Major
Education Specialist
Electrician, Aircraft
Electrician, General
Electric Motor Repairman
Electroplater
Engine Overhaul Mechanic, Aircraft (Designated Engine)
Engineer Stock Man
Fabric Worker, Aircraft
Field Artillery Fire Control Man
Field Lighting Truck Operator
Field Musician
File Clerk
Filter Operator, Water Supply
Financial Typist, Clerk
Finger Printer
Fire Control Instrument Technician
Fire Fighter
First Sergeant
Freight Transportation Clerk
Gas & Oil Man
Guard
Gyro Mechanic, Aircraft
Heavy Artillery Fire Control NCO
Heavy Machine Gunner
Heat-Treater
Hydraulic Mechanic, Aircraft (Designated Type)
Inspector, Aircraft Parts & Accessories
Instructor (Designated Specialty)

Instrument Mechanic, Aircraft
Investigator
Inventory Clerk
Key Punch Operator
Laundry Machine Operator
Legal Clerk
Library Clerk
Link Celestial Navigation
  Training Operator
Link Trainer Instructor
Link Trainer Mechanic
Machine Operator
Machinist
Maintenance Man, General
Materiel Clerk, Aviation
Meat Cutter
Mechanic, Aircraft (Designated
  Type)
Mechanic, Gunner, Aviation
Message Center Chief
Message Center Man
Messenger
Mess Sergeant
Metalsmith, Aviation
Microfilm Technician
Military Policemen
Military Specialty Undetermined
Motor Transport
Multilith or Multigraph
  Operator
Navy Supply Man
Occupational Technician
Office Machine Repairman
  (Designated Machine)
Officer Candidate
Operations Clerk, Aviation
Orderly
Ordnance Stockman
Oxygen & Carbon Dioxide Man
Packer
Painter, Aircraft
Painter, General
Painter, Sign
Painter, Vehicle
Parachute Rigger
Parachute Shop Chief
Parts Clerk, Automotive
Parts Clerk, Ordnance
Passenger Transportation Clerk
Paymaster Clerk
Personnel Clerk
Photographer, Aerial
Photographer, Still
Photographic Darkroom Man
Photographic Laboratory
  Technician
Photographic Service
  Technician
Photographic Stock Man
Photolithographer
Photostat Operator or Blue
  Printer
Plastic Glass Worker
Platoon Sergeant
Plotter, Air Warning
Plumber
Police NCO
Postal Clerk
Post Exchange Man
Process Cameraman
Procurement Clerk
Projectionist, 16mm
Projectionist, 35mm
Projector Operator-repairman
Proofreader
Propeller Mechanic (Designated
  Type
Property NCO
Printer
Publication Man
Quartermaster Supply Basic
Quartermaster Supply Man
Radar Operator (Designated
  Equipment)
Radar Repairman, Airborne
  Search Equipment
Radar Technician (Designated
  Equipment)
Radio Operator, Aerial
Radio Operator, High Speed
Radio Operator, Low Speed
Radio Repairman
Radio Technician, VHF
Radio Telephone Operator
Railway Clerk
Recognition Instructor
Recruiter
Recreation Assistant
Rigger, Aircraft
Sales Clerk
Sewing Machine Operator
Sergeant Major
Sheet Metal Worker
Ship Loading Man
Ship Clerk or Engineer Clerk,
  Aviation
Signal Stock Man
Small Arms Mechanic

Special Assignment
Special Services Assistant
Statistical Clerk
Stenographer
Steward
Stock Clerk
Stock Man, General
Stock Record Clerk
Storage Battery Electrician
Student
Supply Records Clerk
Switchboard Installer, Telephone & Telegraph Dial
Switchboard Operator, Common Battery
Synthetic Devices Mechanic
Synthetic Gunnery Instructor (Designated Type)
Tabulation Machine Operator
Tailor
Telephone Switchboard Operator
Teletype Mechanic
Teletype Operator
Toolroom Keeper
Toxic Gas Handler
Tractor Driver
Traffic Rate Clerk
Training Aids Specialist
Translator (Designated Language)
Truck Driver, Heavy
Truckmaster
Turret Mechanic Aircraft (Designated Type)
Upholsterer
Veterinary Technician
Warehouseman
Watch Repairman
Water Supply Man
Weather Forecaster
Weather Observer
Welder, Acetylene
Welder, Electric Arc
Woodworking Machine Operator
Truck Driver, Light or Chauffeur

TOTAL: 225

Source:

"Brief History," pp. 1-5 (Subject File "Women Marines--History," HistBr, HQMC)

Appendix B: Composition of Women's Reserve: By Education

Did not complete high school	2,608
High school graduate	13,824
1-4 years' college	4,478
Post-graduate work	141
	21,051
Not coded	2,094
TOTAL	23,145 *

* This figure refers to the number of Women Marines who were enlisted during World War II through July 1945.

Source:

Streeter, "History," p. 103.

Appendix C: Composition of Women's Reserve: By State of Residence

California	2,696
New York	2,325
Pennsylvania	1,972
Massachusetts	1,675
Illinois	1,519
Michigan	1,312
Ohio	1,270
Missouri	929
New Jersey	795
Minnesota	745
Wisconsin	716
Washington	599
Iowa	534
Texas	521
Indiana	419
Connecticut	387
Oregon	351
District of Columbia	339
	19,104
Other	4,041
TOTAL	23,145

About 80 percent of the total enrollment of the Marine Corps Women's Reserve during the war came from 17 states and the District of Columbia. Only 20 percent came from the remaining 31 states.

Source:

Streeter, "History," p. 101.

Appendix D: Composition of Women's Reserve: By Age *

From 20 to 24 years	14,300	61.8%
From 25 to 20 years	5,848	25.2%
From 30 to 37 years	2,610	11.3%
Over 38	387	1.7%
	23,145	100.0%

* Records as of 31 July 1945

The majority of Women Reservists enlisted before they were 25 years old, since a considerable number of those in the 25-27 year age group in July 1945 must have been under 25 years when they enlisted. Because the maximum age for enlisted women was 36 years, practically all those over 38 at this date were officers.

Source:

Streeter, "History," p. 102.

Appendix E: Composition of Women's Reserve: By General Classification Test Scores

Range of Test Scores		Total
Under 075		181
075-088		1,045
089-109		8,883
110-129		9,657
130-151		1,220
	Classified	20,986
	Not tested	65
	Not coded	2,094
	TOTAL	23,145

All but 1,226 of 20,986 women were rated 089 or higher.

Source:

Streeter, "History," p. 104.

Appendix F: Key Dates in the History of Women Marines

31 October 1942	--	Secretary of the Navy Frank Knox authorized Marine Corps to create a Women's Reserve and to accept women applicants for commissions and enlistments.
7 November 1942	--	Approval by Commandant, Lieutenant General Thomas Holcomb, of formation of Marine Corps Women's Reserve.
29 January 1943	--	Commissioning of Major Ruth Cheney Streeter as Director, Marine Corps Women's Reserve.
13 February 1943	--	First day that enlistments officially open.
13 March 1943	--	First class of 71 officer candidates enters U. S. Naval Midshipmen's School (WR) at Mount Holyoke, Massachusetts to begin training with the WAVES.
26 March 1943	--	First class of enlisted Women Reserves, numbering 722, begins training at the U. S. Naval Training School (WR) at Hunter College, the Bronx, New York, likewise training with the WAVES.
25 April 1943	--	First class of enlisted women graduated and assigned to active duty. Subsequent classes of approximately 525 women entered every two weeks for courses that averaged about four weeks in length.
4 May 1943	--	First class of officer candidates graduated and report to duty stations. Classes averaged about 70 candidates, began every month, and lasted about eight weeks.
15 July 1943	--	Training for enlisted and candidates having been transferred to Camp Lejeune, North Carolina during past week, instruction for both groups commences here this date. All basic training for Women Reserves, as well as much of the specialty training, is held here throughout the rest of the war.

20 October 1943	--	First candidates' class (i.e., the eighth) composed of meritorious enlisted women begins its training; OCC thereafter comprised largely of former enlisted women.
13 February 1944	--	First Anniversary of Women's Reserve finds organization having grown from four women to nearly 15,000 and well within sight of its recruiting goal: a strength of 18,000 enlisted and 1,000 officers. Original prediction of "more than 30 kinds of jobs" grown to more than 200 different assignments.
27 September 1944	--	Overseas Bill for women in the naval services signed by the President; this allows women naval reservists to serve as volunteers anywhere with Western Hemisphere, including Hawaii and Alaska.
29 January 1945	--	First detachment of five MCWR officers and 160 enlisted women arrives Hawaii for overseas assignment. Later groups of approximately 200 arrive every other week; Hawaii complement eventually totaled approximately 1,000 women.
13 February 1945	--	Second Anniversary of Women Reserves celebrated with dances, birthday cakes, special religious services, and battalion reviews. The women numbered from one-third to one-half of the post troops at many Marine duty stations.
7 May 1945	--	V-E Day. Recruiting for Women's Reserve limited to replacements for normal attrition.
2 September 1945	--	V-J Day. All recruiting stopped, and plans made for gradual demobilization of Women's Reserve.
13 February 1946	--	Some 1,700 Women Reserves marched smartly in review before Commandant, General Alexander A. Vandegrift, in ceremonies at Washington, D, C. marking the Third Anniversary of the Women's Reserve.
7 June 1946	--	Approval by the Commandant of Marine Corps Women's Reserve Policy Board recommendation for retention of small number of women on duty to serve as trained nucleus for possible mobilization emergencies.

1 September 1946	--	Original terminal date set for Women's Reserve. All WR units disbanded and most of women returned to civilian life.
12 June 1948	--	Passage of Women's Armed Services Integration Act established Women Marines as a permanent part of regular component of Marine Corps, as well as permanent reserve status.
4 November 1948	--	First group of three wartime WR officers sworn into the regular Marine Corps.
10 November 1948	--	First group of eight World War II enlisted women similarly sworn into the regular Marine Corps by the Commandant.

Source:

Recapitulation of facts in this monograph.

MRS. RUTH CHENEY STREETER
FORMER DIRECTOR OF MARINE CORPS WOMEN'S RESERVE

Colonel Ruth Cheney Streeter was the first Director of the United States Marine Corps Women's Reserve. She earned the Legion of Merit for "outstanding services" during World War II and served from the time the Women's Reserve was activated on February 13, 1943, until December 7, 1945, when she resigned her commission.

The colonel was awarded the Legion of Merit on February 4, 1946. The accompanying citation states in part: "Exercising judgment, initiative and ability, Colonel Streeter rendered distinctive service in directing the planning and organization of the Women's Reserve of the Marine Corps and skilfully integrating women into the basic structure of the Corps, carefully selected, trained and properly assigned them as replacements for men in shore establishments."

Born October 2, 1895 at Brookline, Mass., Colonel Streeter attended schools abroad and graduated from Bryn Mawr College at Bryn Mawr, Pa., in 1918. During the depression years following 1930 she worked in public health and welfare, unemployment relief and old-age assistance in her home state of New Jersey. She was one-time President of the Welfare Board in Morris County, N. J. She also served as a member of the New Jersey State Relief Council, New Jersey Commission of Inter-State Cooperation, and New Jersey Board of Children's Guardians.

Long interested in aviation, the colonel completed a course in aeronautics at New York University and served as adjutant of Group 221, Civil Air Patrol. She learned to fly in 1940 and in 1941 became the only woman member of the Committee on Aviation of the New Jersey Defense Council. The same year she also acted as chairman of the Citizens' Committee for Army and Navy, Inc., for Fort Dix, N. J. She received her commercial pilot's license in April 1942.

Colonel Streeter was the first woman to hold the rank of major in the Marine Corps. She was appointed to that rank on January 29, 1943. She was promoted to lieutenant colonel on November 22, 1943, and to the rank of colonel on February 1, 1944

COLONEL KATHERINE A. TOWLE, USMC

Colonel Katherine A. Towle took office as Director of Women Marines on November 4, 1948, after she became one of the first three women officers in the regular Marine Corps.

General Clifton B. Cates, Commandant of the Marine Corps, administered the oath of office to Colonel Towle, who also served as Director of the U. S. Marine Corps Women's Reserve from December 7, 1945 to June 12, 1946.

Colonel Towle was born in Towle, California April 30, 1898, the daughter of the last George Gould Towle and Katherine Meister Towle.

She was graduated from the University of California at Berkeley in May, 1920 with an A. B. degree, later receiving the M. A. degree in political science from that school. In addition she has studied at Columbia University in New York City.. From 1929 until 1932 she was headmistress of the Miss Ranson and Miss Bridges School for Girls at Piedmont, California.

When called to active duty simultaneously with the receipt of her commission as captain in the U. S. Marine Corps Women's Reserve on February 25, 1943, she was employed as Assistant to the Manager, University of California at Berkeley. She was one of the first to be commissioned in that component of the Marine Corps, and holds another "first" title with her appointment as the Director of Women Marines under the terms of the Women's Armed Services Integration Act passed by the 80th Congress and signed by President Truman in June, 1948.

In early March, 1943 she was ordered direct from civilian life to Marine Corps Headquarters in Washington, D. C. Later that month she was ordered to the Marine Detachment Naval Training School (Women's Reserve), Hunter College, New York City, as the senior woman officer of the detachment.

In May of the same year she was detached from Hunter College ordered to temporary duty in Washington, and in early June was assigned to the special staff of the Commanding General, Camp Lejeune, New River, North Carolina as "Assistant for Women's Reserve," with the opening of the Women's Reserve Training Center there.

While serving in that capacity in February, 1944, she was advanced to the rank of major.

96

Her next duty assignment, beginning in September , 1944, was in Marine Corps Headquarters as Assistant Director of the U. S. Marine Corps Women's Reserve. In March, 1945, she was promoted to the rank of lieutenant colonel, and with the resignation of the Director of the Women's Reserve, she became the second Director on December 7, 1945. Her advancement to the rank of colonel came simultaneously with her appointment by General A. A. Vandegrift, then Commandant of the Marine Corps.

Colonel Towle was awarded a Letter of Commendation, with Ribbon, in March, 1946, for "meritorious service during the entire period of the growth and development of the United States Marine Corps Women's Reserve...." Other decorations include the American Campaign Medal and the World War II Victory Medal.

On June 12, 1946 Colonel Towle relinquished her position as Director of the Women's Reserve and returned to the University of California at Berkeley following her release from active service on August 18, 1946. From that time until she reported to the Commanding General, Department of Pacific, San Francisco, for active duty on September 23, 1948, she was assistant dean of women at that university.

The colonel reported for duty at Marine Corps Headquarters October 18, 1948.

- USMC -

Prepared by Division of Public Information,
Headquarters Marine Corps, February 1949

Made in the USA
Monee, IL
05 November 2026

67385030R10057